IMAGINING THE SPHERES
By Dr Steven Hobbs

Including work by members of the International Association of Astronomical Artists
and pre space-age artist David A. Hardy

ISBN: 978-0-6484476-0-3

(Preceding page) A feeble sun struggles to break through the eternal smog of Saturn's moon, Titan.
(Facing page) Vallis Marineris, the largest canyon in the solar system, splits the crust of Mars. Vallis Marineris is so big that it occupies one quarter of the planet's circumference.

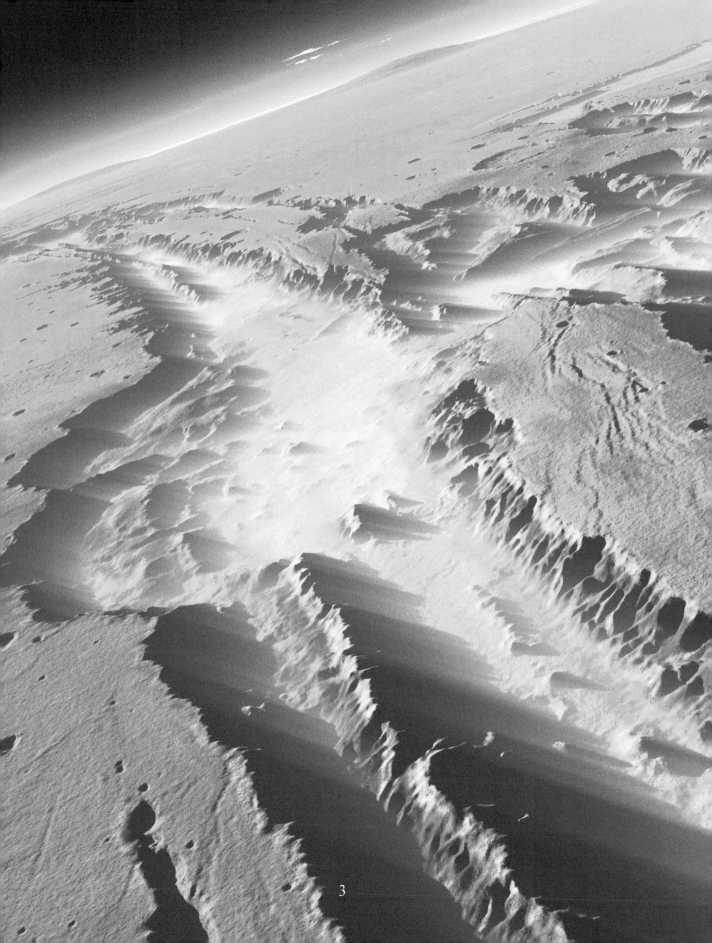

3

FOREWORD
By Alastair Reynolds

Science and art, art and science. I've never been too keen to draw a sharp line between the two, and this beautiful and compendious book strikes me as a fine celebration of that decidedly porous interface. Long before our probes and instruments glimpsed these alien vistas, our imaginations had already visited them. Those visions of the pre-space age might have been outmoded – in some cases wildly so – but in retrospect the misses are as illuminating as the hits, offering a fascinating glimpse into the state of scientific knowledge and speculation before the advent of close-up observations.

There's a remarkable juxtaposition early in the book. In a painting by David Hardy, based on the British Interplanetary Society's ideas for a Moon expedition, a lunar astronaut lifts a hand to their helmet while behind them a gently rounded lander sits amid craggy, angular rock formations. In the corresponding photo from the Apollo era, a lunar astronaut lifts a hand to their helmet while behind them a craggy, angular lander sits amid gently rounded rock formations.

The idea of a Moon dominated by towering, sharp-edged pinnacles persisted well into the early space age. There were valid reasons to think such a landscape might be the case: the Moon won't have been subjected to terrestrial weathering processes, and the shadows cast by the terminator (the line between the lit and unlit faces) look incredibly sharp, suggesting cliff-like contours. Even now, it's hard not to look through a small telescope and feel that the Moon is sharp-textured, chiselled. It just looks that way. Science tells us otherwise, though, and our imaginations must re-calibrate for this new reality.

The David Hardy connection is particularly pertinent to me because he has always moved freely between the scientific and artistic realms. Indeed, a picture by Hardy may be one of the earliest pieces of astronomical art I ever "owned". It came to me by way of an album sleeve, wrapped around a record that was given to me somewhere around my sixth or seventh birthday. The recording, featuring the London Philharmonic Orchestra, included "Theme from 2001" (Strauss's Also Spracht Zarathustra), as well as a number of planet-inspired compositions by the somewhat obscure Wilford Holcombe. The synthesizer-led music did powerful things to my imagination, drawing evocative mental images. I still can't listen to Holcombe's Venus without envisioning a lone, space-suited figure crossing a pitiless sun-blasted plain, seeking the distant sanctuary of a cave or base. As for Hardy's stirring cover illustration, which depicts two astronauts gambolling around

one of the Martian moons in low gravity: I remember being very slightly disappointed that this scene didn't actually feature in Kubrick's film. Ah, the naivety of youth!

Hardy is well represented here – his career has been gratifyingly long and wide-ranging – but the work of numerous other talented space artists is also featured, including such luminaries as Ron Miller and Bill Hartmann, as well as some fine contributions from Steven Hobbs himself. Given that astronomy has a less than spectacular record when it comes to gender parity, it's perhaps not too surprising that there are only a couple of images from female space artists included. That said, a quarter of planetary scientists (as of 2017) are women, even if they are under-represented at the highest levels of their profession, and we may hope that future equivalents of this book will skew toward a more equal gender balance. We are, after all, talking about a field in which some of the key findings have been made by women, such as (to pick only one example) the first discovery of active volcanism beyond our own Earth.

Setting aside individual attributions, the art in this book is gorgeous and inspiring. Even when science has invalidated the thinking behind a particular image, aesthetic enjoyment is still to be had. It's doubtful that we'll ever see Saturn suspended in the skies of Titan, as depicted in a couple of paintings here, but who doesn't wish that were the case? The best art endures, regardless, and who knows, nature may yet deign to throw us a surprise or two when our probes (and eyes) get to spend serious time in these landscapes and environments. Artists, meanwhile, will be motivated by observations and predictions we can't yet envisage.

Thus the great dance between art and science will continue, and nor will it end at the frontiers of our own solar system. Telescopes and space-based observatories have revealed to us the existence of exoplanets, and in some cases their sizes, densities and surface conditions may be estimated with some confidence. But we are still a long way from the direct imaging of an exoplanet's face, and we may be centuries from getting close-up and personal with these distant alien worlds. They really are a quite staggeringly long way away.

In the meantime, art and imagination – guided by science, certainly – has nearly free rein to take us there ahead of time, unconstrained by distance or budgetary cycles. Artists have indeed already begun colonising these distant vistas and extreme environments, while the discovery of new exoplanets shows no sign of abating. The artists of the future may work in traditional media or digital (indeed, some of these artists may even be digital themselves) but one thing is certain: their visions - as beautiful, weird, desolate or chilling as they may turn out to be – will constitute one small but not insignificant step in making us want to go out there for real. And we will, taking our paintboxes – metaphorical or actual – with us.

But even then, we'll really only just be starting out.

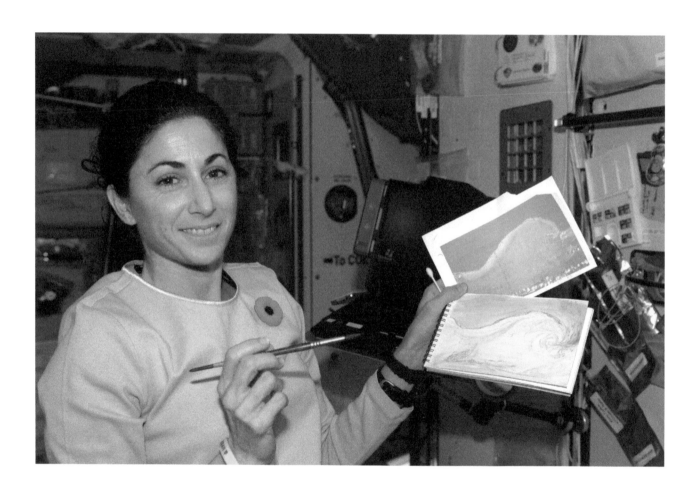

(Above) Pioneering astronaut Nicole Stott painting the first watercolours in space from the International Space Station. Nicole's paintings promote the amazing work being done every day in space to improve life right here on Earth(Nicole Stott).

CONTENTS

(Above) A pre space-age painting of the surface of our Moon by Lucien Radaux (Public domain).

INTRODUCTION

Planet Earth is our home. We once thought it to be the physical centre of the universe, around which everything revolved, until astronomical observations revealed the Earth and our fellow planets indeed move around the Sun. Centuries-old beliefs and assumptions resisted the "new" truth, but eventually the field of astronomy emerged as a true science. The efforts to learn about space and our neighbouring worlds expanded the limits of human imagination.

For thousands of years, the ancients observed the "wandering stars" in the sky, amazed at their ability to move in the unchanging heavens, even assigning them god-like qualities. During the Renaissance, the Moon and planets were assumed to be Earth-like. Oceans lapped the grey shorelines of the Moon, and sea monsters inhabited the planet-wide ocean of Saturn. The idea that planets would be made of poisonous gas, or moons made of fire, was as remote as the planets themselves.

As observing the night sky evolved from superstition to a measured science, our neighbours became distinctly less Earth-like and more and more alien. The advent of the space age sped this process up immensely. As our space probes visited planet after planet, millennia of philosophy and assumed understanding were transformed, literally overnight.

And like every other great human endeavour, human imagination helped inspire the space age. Artwork depicting scenes of standing on the surface of other worlds set generations of astronomers and scientists on a path to design missions to go there. Artists have visualised spacecraft concepts before being built and directly assisted in refining their designs. Artistic renderings changed the design of the Beagle 2 Mars lander, when an artist discovered components could not fit as designed. Even entertaining films of enchanting Venusians and Martian queens helped inspire many of the people who architected the Space Race.

It is these scientists and engineers who turned the dreams of exploration into practical reality. Although pioneering giants such as James Van Allen, Sergei Korolev, or Wernher von Braun rarely became household names, they and many others have worked hard to conquer environments that are beyond human ability to endure. Our understanding of who we are in the cosmos, as well as technological spin-offs from space exploration, are direct outcomes from their work.

As we approach the 50th anniversary of Apollo 11, there is a dichotomy in space exploration. Despite plans of return missions and accessible spaceflight promised for decades, no human being has risen higher than 600 nautical miles above Earth since 1972. Since then, many of our great space pioneers have fallen. Less than half of the 12 men who walked on another world remain among the living. Soon, no-one will be able to tell us what it was like to ride the largest rocket ever built, or cover the whole Earth, and everyone who ever lived, with their thumb.

On the other hand, within a human lifetime, explorers of a different kind have flown past, landed on or smashed into our neighbouring worlds. Through our robotic proxies of iron and glass, we have visited every single planet in the Solar System. People born after 1997 have never known a Mars that was not continuously observed.

Artists have followed the unfolding space age all the way. Even now space artists take advantage of volumes of data returned from the latest space missions to create new visualisations of the Solar System.

Only a few space artists were alive and working before the dawn of space ships. Their careers spanned the transformation of the Moon and planets from distant unknowns to scientific reality. Artists such as Ludek Pesek, Chesley Bonestell, and David A. Hardy started out painting canals on Mars and jungles on Venus, and saw the missions from the 1960s and 70s alter forever our ideas of what the Solar System really looks like.

Thanks to the artists, scientists, and engineers of the world's space agencies, people living in the first decades of the 21st century now see the Moon and planets as real places. Space science and exploration pops up in casual conversation, and many people imagine travelling to these strange landscapes in a far more personal way than ever before.

This book is a journey in words and pictures of how the Space Age changed our understanding of our neighbouring worlds. It is a story of human imagination, endurance, and ingenuity, and how the quest for the answers of tomorrow pushed people to achieve the once impossible, driving human dreams ever forward.

(Above) The crescent Earth as seen from the far side of the Moon (Don Dixon).

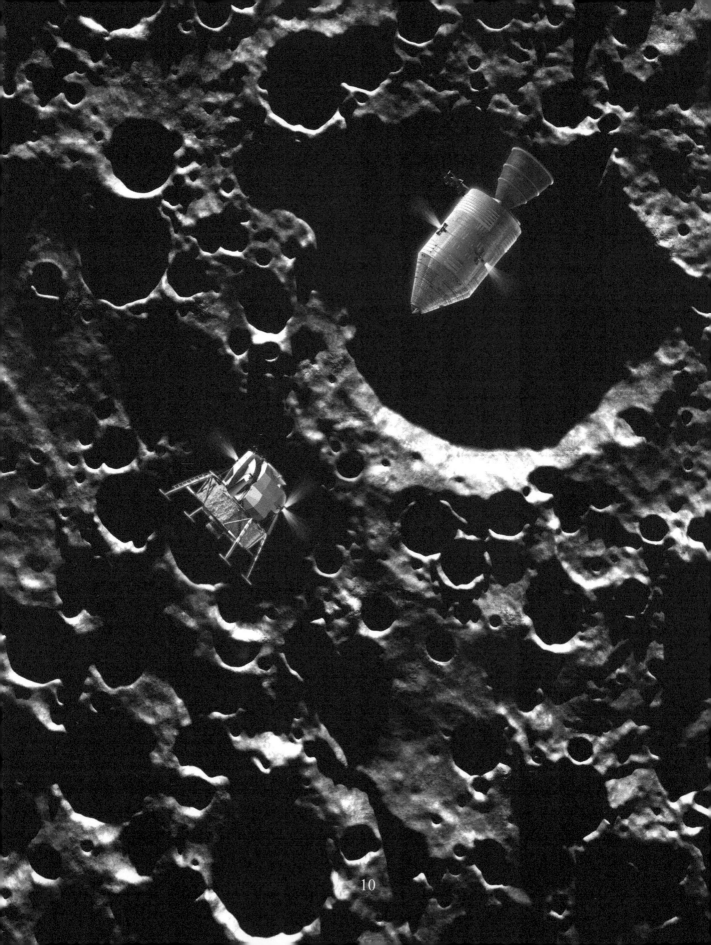

MOON

"That's one small step for [a] man, one giant leap for mankind."

This is perhaps the most daring, if not famous, phrase ever spoken in the history of humanity. With these words, Commander Neil Armstrong stepped off the footpad of his spacecraft and touched another world. The Apollo Moon landings are arguably one of the defining moments of the 20th Century. Born from the ashes of World War II, in many ways, they represent the pinnacle of human technological prowess, exploration and adventure. In the half-century that has passed, many have yearned for a return to the days where, for a brief period, humanity pulled together to achieve something many thought impossible. Such a return has seemed just beyond our grasp. Although our technology has advanced massively, our desire to undertake risks and massive projects has diminished.

Following the Sun, our Moon is the most obvious thing in our sky. References to the Moon have appeared in most cultures and religions. The Moon's journey through the skies has been used for millennia to mark seasons and govern crop growing and harvest. Many superstitions and unearthly fears have come from our celestial neighbour, with lunar madness and even werewolves brought on by our nearest celestial neighbour. It was a brave peasant who journeyed through the wilderness alone under a full Moon.

Before the telescope, the popular consensus followed the Aristotelian view that all of creation, including the Moon, revolved around the Earth. The Moon had been seen as a perfect celestial sphere, free from the blemishes of the corrupted Earth. The light and dark patches were viewed as reflections of Earth below. This view was seriously challenged when Italian scientist Galileo Galilei pointed a telescope at the Moon and discovered patterns of light and shadow:

"Let me speak first of the surface of the Moon... I distinguish two parts in it, which I call respectively the brighter and the darker. The brighter part seems to surround and pervade the whole hemisphere; but the darker part, like a sort of cloud, discolours the Moon's surface and makes it appear covered with spots. These spots have never been observed by any one before me; and from my observations of them, often repeated... I feel sure that the surface of the Moon is not perfectly smooth, free from inequalities and exactly spherical, as a large school of philosophers considers with regard to the Moon and the other heavenly bodies, but that, on the contrary, it is full of inequalities, uneven, full of hollows and protuberances, just like the surface of the Earth itself, which is varied everywhere by lofty mountains and deep valleys."

For the first time, the Moon was revealed to be not a reflective sphere, but with topography. Suddenly, the Moon became a world in its own right. Over the next two centuries, as telescopes improved with the industrial revolution, the Moon was revealing itself to be quite different from the Earth. Firstly, astronomers saw no evidence for oceans, streams or clouds familiar to our home world. In the 1800's, Irish professor Sir Robert Stawell Ball examined the Moon:

"The Lunar landscapes are excessively weird and rugged. They always remind us of sterile deserts, and we cannot fail to notice the absence of grassy plains or green forests such as we are familiar with on our globe." The dark grey areas, long thought to be seas by the ancients, lay unmoving and still.

Past astronomers searched for an atmosphere by watching the Moon pass in front of a star. If a gaseous mantle were present, a gradual dimming would be expected. Instead, star after star simply winked out like a blown-out candle. For most, the Moon was turning out to be a sterile desert, where lack of wind and rain would preserve the Moon's features for aeons, and life would not be possible.

Others began noticing that the mountain ranges on the Moon were round. In 1847, European astronomer Thomas Dick remarked in Celestial Scenery: *"... circular ranges ... appear on almost every part of the moon's surface, particularly in its southern regions. This is one of the grand peculiarities of the lunar ranges to which we have nothing similar in our terrestrial arrangements. A plain, and sometimes a large cavity, is surrounded*

(Facing page) The Apollo Lunar Module separates from the Command Module to begin a perilous descent to the Lunar surface. Only six times have such spacecraft, and any of humanity, ever visited the Moon.

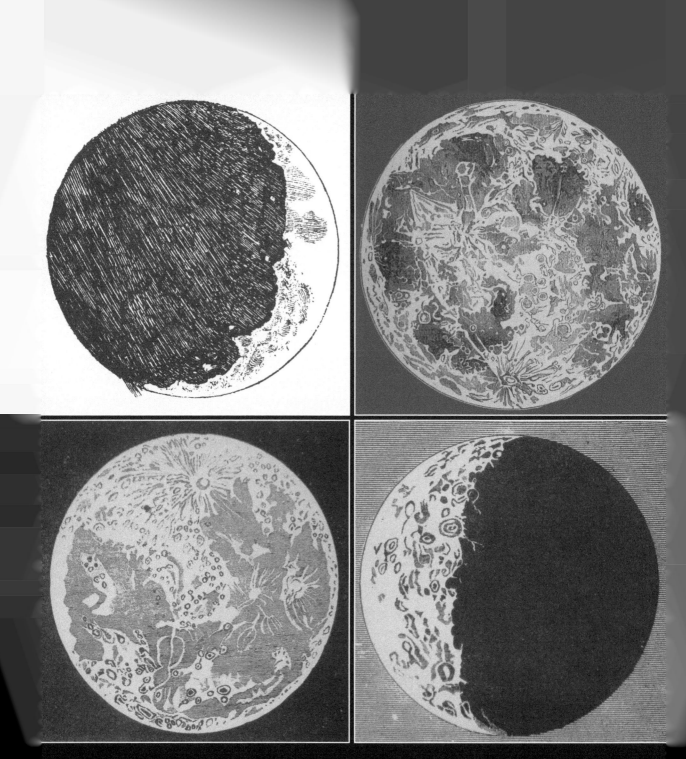

(Above) From Galileo's first sketches of the Moon (top left), astronomers realized that our neighbour was a world with mountains and valleys. Pre-photographic sketches from the 19th Century (top right, above left, above right) identified dark grey seas, circular craters and crater rays (Public Domain). (Opposite) The long, raking shadows near the Moon's terminator, as shown in this 19th Century sketch, fooled many astronomers into thinking that the Moon's Lunar Mountains were sharp, snowy peaks (Public Domain).

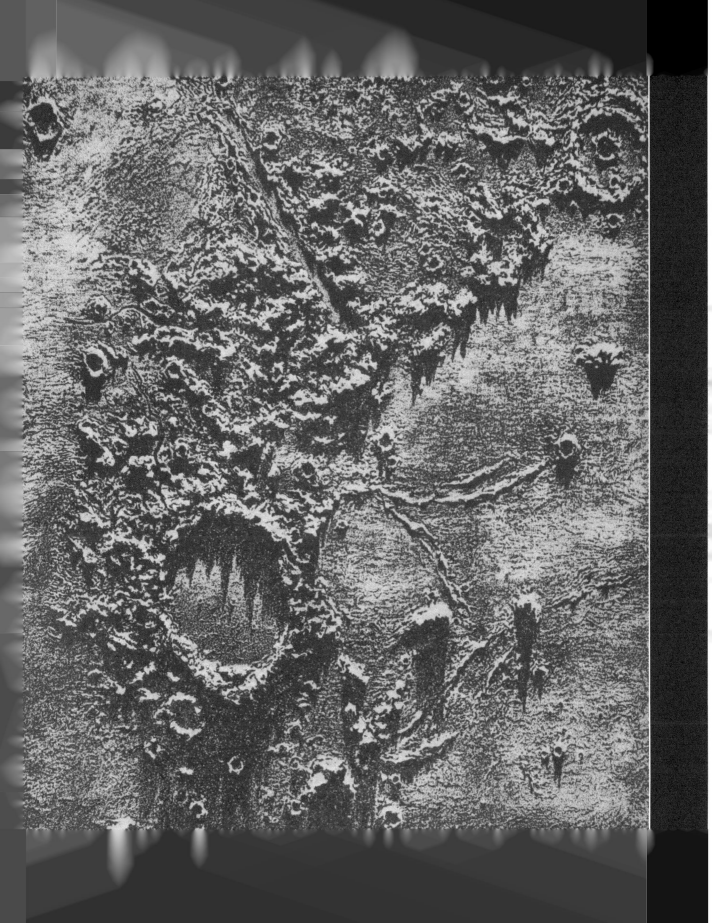

(Above) A Victorian-era view of the cratered lunar surface. Most 19th century astronomers saw the Moon as a world covered in volcanos (Public Domain).

with a circular ridge of mountains, which encompasses it like a mighty rampart." Dick was describing craters. Astronomers' telescopes discovered hundreds of these circular impressions riddling the Moon's surface like a pock-marked mess. The closest thing Earth had to these features came from volcanic eruptions, and Victorian-era scientists imagined an ancient Moon wracked with the 'Fires of Hell'. In 1895, Sir Robert Stawell Ball wrote: *"It seems certain that in ancient days great volcanoes abounded on our satellite…The volcanoes must then have been raging on the moon with a fury altogether unknown in any active volcanoes which this Earth can now show."* For others studying the Moon around this time, the volcano hypothesis began to fall out of favour. What caused the bright rays streaking from some of the large craters? Also, why were there strings of smaller craters lying nearby larger ones? Could they be related?

To an observer, the scene must have looked almost comical: a Victorian-era American gentleman, shooting rounds into perfectly good mounds of clay. Perhaps inspired by the discovery of Meteor Crater in Arizona, geologist Grove Karl Gilbert performed such experiments as he pursued evidence for an alternative origin for Lunar craters. Studying the impacts left from his rifle bullets on targets, he began to imagine the Moon as a target of its own in a cosmic shooting gallery. Impacts from space rocks, raining down over the eons, would blast out larger versions of the round holes seen in his experiments. Debris from these collisions would rain down to form the bright rays seen in the Moon's larger craters, or even create chains of secondary craters. This view had also been proposed in the 1870s by English astronomer Richard A. Proctor. Analysing the frequency of which meteors would have struck the Moon in its history, he wrote:

"The evidence in favour of the meteoric theory of the small craters is much stronger than I at first supposed, the difficulty of forming any other plausible theory much greater." Perhaps taking a jab at the volcano theorists, Proctor continued: *"I may even go so far as to say it would be a problem of extreme*

14

(Above) A Lunar Reconnaissance orbiter image of a lunar crater. Spacecraft images of the Moon showed that our neighbor's craters were formed by meteor impacts (NASA).

difficulty to show how a body formed like the Moon, exposed to... the same enormous time-intervals, could fail to show such markings as actually exist on the moon."

By the start of the space age, astronomical Lunar exploration had just about reached its limit. Earth's turbulent atmosphere permanently hid features smaller than a city block from view, and only half the Moon would ever be visible from Earth. The Lunar far side could well have had a bustling civilisation on it and the people of Earth would never know. The development of reliable rockets during World War II would soon give humanity its chance to get a closer look.

As far as targets of space exploration go, the Moon makes an ideal choice. Its proximity to our

home world, compared to anything else in the solar system, allowed it to be reached with modest amounts of rocket power. Despite being the first to take advantage of German V-2 missile technology after World War II, America was taken by complete surprise when Sputnik, humanity's first artificial space satellite, was launched from Russia, in 1957. This single event began the space race between the two superpowers, with many seeing it as a contest between American democracy and Soviet communism. The paranoia of the cold war was at its highest, and neither nation wanted to be seen to be second to the other in technological prowess or political prestige. This competition may be viewed harshly, but some have argued that the manned Moon landings and the accelerated technological growth that came with them would not have happened without it.

Humiliated by not being the first to reach Earth orbit, America tried desperately to beat Russia to the Moon with the US Air Force's Pioneer series probes, starting with Pioneer 1 which was launched a year after Russia's Sputnik success. Pioneer 1 was to enter Lunar orbit, and among other tasks, photograph the Moon's far side. Unfortunately, Pioneer 1's booster failed to impart enough energy for it to reach the Moon and it, as

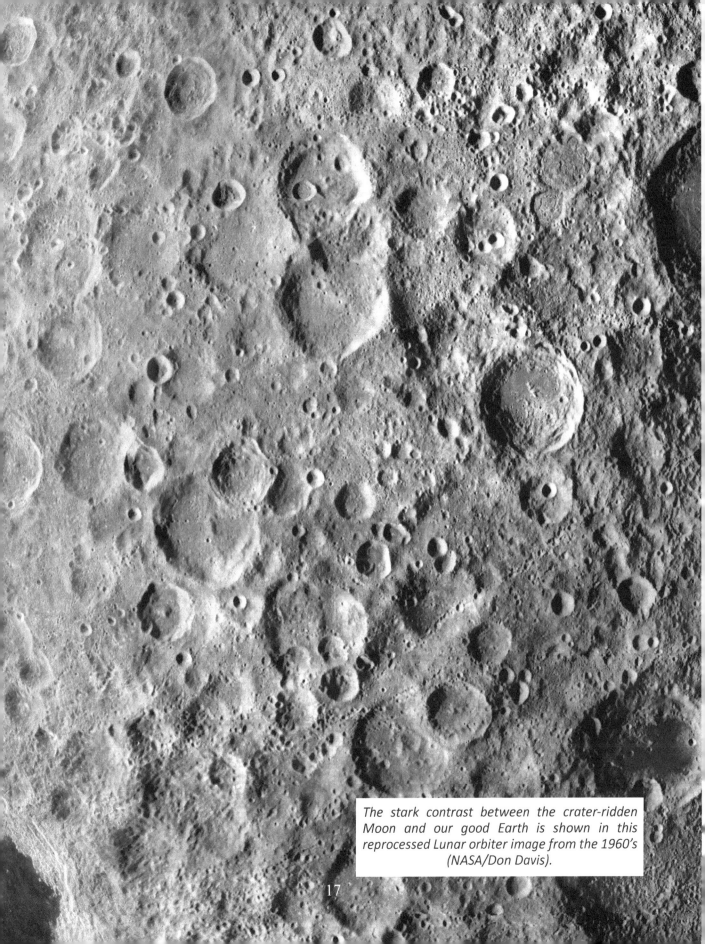

The stark contrast between the crater-ridden Moon and our good Earth is shown in this reprocessed Lunar orbiter image from the 1960's (NASA/Don Davis).

well as two later Pioneers, failed to break free from Earth. Russia again triumphed when the beach ball-sized Luna 1 flew within 6,000 kilometres of the Moon on 4 January 1959. Luna 2 followed, becoming the first artificial object to impact the Moon in September of 1959. As if to press the point, the Soviet Premier Nikita Khrushchev presented a copy of Luna 2's pennant to the then US President Eisenhower during his visit to the USA.

America's new space agency, the National Aeronautics and Space Agency (NASA), was reeling at this point. It received a further blow on Sputnik's second anniversary, when Luna 3 was sent around the Moon's far side. More sophisticated than anything that had reached the Moon before it, Luna 3 photographed the lunar far side with film cameras. As the probe's elliptical orbit brought it closer to Earth, a photoelectric multiplier converted the image to an electrical signal that was laboriously transmitted to mission controllers on the ground. Although the process stopped before the crude images could be refined, 70 percent of the Moon's far side was mapped. The Luna 3 scientists received a case of expensive champagne from a rich French eccentric that had promised the wine to whomever could show him the Moon's far side.

The US responded to the Soviet challenge: *"I believe that this nation should commit itself to achieving the goal, before this decade is out, of landing a man on the moon and returning him safely to the earth. No single space project in this period will be more impressive to mankind, or more important for the long-range exploration of space."* These words, spoken by a young President John F. Kennedy in 1961, set off a chain of events that would lead to the greatest technological undertaking in human history. In nine short years, after just 15 minutes of human spaceflight experience, the US would have to invent, test, launch and land vehicles that tamed the energy equivalent of a thermonuclear weapon, and land on a body that had just recently emerged from the realms of science fiction. At its peak, the Apollo project would consume nearly 5 percent of the US federal budget, employ almost half a million people worldwide, and develop technologies that paved the way for the electronic revolution seen in recent years. Kennedy also said: *"... we shall send to the moon, 240,000 miles away from the control station in Houston, a giant rocket more than 300 feet tall..."* The need to send tonnes of mass required to support men on the Moon would lead to the creation of the largest ever rocket in history.

(Above) A segment of the first ever image from the Moon's surface taken by Luna 9. Dust and small pebbles can be seen (NASA).

(Above) An image of the US lander Surveyor 3's footpad. The lander bounced twice on the ground before finally coming to rest (NASA).

"... made of new metal alloys, some of which have not yet been invented..." The required metallurgical experience simply didn't exist. NASA would have to start from scratch. *"... on an untried mission, to an unknown celestial body..."* In 1961, the surface of the Moon remained a mystery. Nothing smaller than a football field could be resolved. Were there house-sized boulders ready to smash a fragile lunar lander, or would the first men sink in a sea of fine dust?

"... then return it safely to earth, re-entering the atmosphere at speeds of over 25,000 miles per hour, causing heat about half that of the temperature of the sun--almost as hot as it is here today--and do all this, and do it right, and do it first before this decade is out--then we must be bold."

As NASA tooled up to develop sophisticated spacecraft, trial changing orbits and docking in space, robots were hastily developed to perform a thorough reconnaissance of the Lunar surface. David A. Hardy, whose space art career spanned the space race, was inspired by Kennedy's speech: *"These words were what I had been hoping to hear for just about ten years, since my first published painting of a Moon landing was in 1952! Of course, it was all rather different in the reality than we had imagined in the fifties; no gleaming, streamlined, winged spaceships landing on the Moon complete. No space stations in Earth-orbit to pave the way. And when we left there in 1972 it wasn't with the intention of building a lunar base — we would never go back, and have not to the present day... Even so, seeing those blurry, black-and-white images on TV in 1969 was magical, and the fulfilment of a dream."*

The 1960's started full of promise for both sides of the Iron Curtain. The former Soviet Union had begun to design spacecraft that would use retrorockets and air bags to soft land a probe on the Moon. Meanwhile, the US had commenced on an ambitious lunar program called Ranger that would eventually spawn the highly successful Mariner and Voyager spacecraft to the planets.

The US Ranger program started with five spacecraft, consisting of two parts. A mother ship equipped with cameras, a number of experiments and a communications dish carrying a smaller balsawood-shrouded seismometer experiment that would separate and hard land on the Moon. The mother craft would photograph the lunar surface until it crashed on the Moon, kamikaze style. Part of Ranger's mandate was to determine whether the Moon's surface would be strong enough to support the weight of a manned spacecraft. Some

(Above) Surveyor photographs its own shadow on the Moon. Colour photography of the mostly grey Moon had limited reward (NASA).

(Above) NASA's Lunar Orbiter spacecraft circles the Moon, mapping landing sites for future Apollo missions (Mark Pestana).

19

scientists believed that eons of meteor bombardment had led to the creation of vast seas of lunar dust that would swallow men and spacecraft whole.

Despite being hailed as 'brave attempts' by the press of that time, all five Rangers failed. Only Ranger 4 actually made it to the Moon, suffering the equivalent of a fatal stroke in its communications equipment and falling silently to its dusty grave. The ever-watchful Soviets promptly chided America by mentioning that the pennant left behind by Luna 2's Moon impact two years ago was getting lonely. It is somewhat ironic that America's Mariner 2, spawned from the Ranger project, had managed to journey all the way to Venus, a target many times further away, two years before the nation achieved success on the Moon. The Ranger project manager was promptly sacked and the spacecraft were stripped of most science experiments to virtually become a meteor with cameras.

Although Ranger 6 failed, Ranger 7 successfully impacted on the Moon on 30 July 1964. For the two minutes before its destruction it returned 4300 of the clearest images taken of the Moon at the time amidst shouts and wild cheering among an audience of overworked, overtired mission controllers. Following another successful mission with Ranger 8, the Jet Propulsion Laboratory in Pasadena, California took a big risk with their last spacecraft and pushed for Ranger 9's mission to be viewed live on major television networks. Children were glued to their television sets as Ranger 9 raced to its demise in the crater Alphonsus. Finally, America had triumphed on the Moon.

The Russians also started experiencing significant problems with their lunar program. During what the Soviets called 'a series of experiments' in the early 1960's, their first few soft landers failed to launch, missed the Moon entirely or crashed, their delicate electronics scattering in the lunar dust. Finally, in 1966, Luna 9's retrorockets fired, slowing its downward plunge to the lunar surface. A hinged sensor arm touched the ground, causing an airbag-shrouded capsule to drop off its supporting lander craft and bounce in the weak lunar gravity. Their job done, the airbags split open and flew away, leaving the weighted capsule to right itself and unfurl its petals, revealing a panoramic TV camera. After an anxious four-minute wait, the Soviet scientists received Luna 9's transmissions right on time, signalling the first ever probe to successfully soft land on the Moon. Luna 9 survived on the Moon for three days, transmitting panoramas consisting of dust and rocks stretching to a horizon that, due to the Moon's smaller size, appeared much closer than on Earth.

The West, defeated yet again in the race to the Moon, tried desperately to rob the Soviets of some of their glory. Luna 9's image transmissions were received by the Jodrell Bank tracking station, at the time an important radio receiver facility in England. A British radio astronomer allegedly leaked the intercepted images to the Western press. The very next day, Luna 9's pictures appeared in the Daily Express, beating the Soviet newspaper, Pravda, to the punch and infuriating the Russians. Despite this, Luna 9 proved that a spacecraft wouldn't sink into a sea of lunar dust, making it possible for future manned landings.

Pravda was able to make new headlines when, two months later, the Communist Party's anthem was played from Lunar orbit by the first spacecraft ever to do so, Luna 10. Luna 10 demonstrated that a spacecraft could achieve a stable lunar orbit and, by using an on-board Geiger counter, determined that radiation levels were well within human tolerances. These findings had direct implications for manned lunar spaceflight and sent NASA into a panic.

NASA's answer to the Soviet Luna lander series was the Surveyor program, which was rushed forward to prepare for the Apollo Moon landings. Between 1966 and 1968, these sophisticated spacecraft used a solid rocket engine and three smaller control engines to slow it down from 9,344 kilometres per hour to 5 kilometres per hour. Landing gear designed from lunar data gathered by the Ranger program cushioned Surveyor's final four-metre free-fall to the ground. Science packages of increasing sophistication enabled the Surveyors to return the first colour photographs of the lunar landscape and sample the lunar soil directly with scoops and spectrometers. The Surveyors were complemented by five lunar orbiters, each containing 75 kilograms worth of cameras that provided from lunar orbit images so detailed they are still used today. So advanced were the Surveyor landers that if they had landed on the Moon before Luna 9 there would have been little original lunar science left for the simple-minded Soviet probes to do. As it was, Surveyor 1 successfully landed on the Ocean of Storms barely three and a half months after its Soviet competitor. For the next six weeks, including a two week 'sleep period' over the long lunar night, Surveyor 1's filtered cameras sent back over 11,000 colour

Doug Forrest SEP. 2012

(Above) Commander Neil Armstrong, soon to be the first human to walk on the Moon, adjusts his spacesuit before walking out to the Apollo 11 rocket. As today, pre-flight suit checks were vital before any spaceflight in

pictures, providing valuable reference material for the Apollo lunar landings being planned.

Thanks to faulty surface radar readings on approach, Surveyor 3 bounced twice on the Moon like an errant beach ball before finally coming to rest. Its soil scoop dug into the lunar regolith, pioneering a procedure that would be repeated four decades later by the Phoenix probe on Mars. Professor Jim Head, who was a geologist working on the Apollo program, describes the effect of the Ranger and Surveyor programs on understanding the Moon: *"No one had any real idea about the nature of the surface. [Austrian astrophysicist] Tommy Gold was convinced that the spacecraft and astronauts would sink into a sea of dust. These missions unequivocally put that idea to rest in everyone's mind (but Tommy's) and paved the way for our traverse and rover planning for the Apollo missions."* Surveyor confirmed that lunar soil would indeed hold an astronaut and their lander, having the consistency of wet beach sand.

Surveyors 5 and 6 followed Surveyor 3 to successful landings. Surveyor 5 measured the composition of the lunar soil with an alpha spectrometer, and Surveyor 6, following completion of its primary mission, was relaunched to explore a new area, three metres from its initial landing site. Surveyor 7, the last and most sophisticated in the series, was allowed to visit a more interesting part of the Moon than its Apollo proving-ground predecessors. To the delight of lunar geologists, it made a flawless landing among the gully and boulder-ridden region of the crater Tycho. A faulty alpha backscatter probe had to be helped to the ground by the soil scoop, but Surveyor 7 was in all other respects highly successful and managed to return 21,000 images of the lunar highlands, along with soil chemical analysis data.

The Surveyor and Lunar Orbiter series represented the crowning glory of automated Moon exploration by NASA. These missions, in their design and control, set the benchmark for future unmanned space exploration. Later spacecraft to the Moon and other planets followed similar procedures in design, testing and mission control. All but two of the spacecraft had succeeded in their mission; only Surveyors 2 and 4 missed their marks. Among the NASA spacecraft greatest assets were their cameras. Although beaten in the race to return lunar imagery by the Soviets, NASA pictures were of higher resolution by orders of magnitude than their Russian counterparts. The Soviets in turn developed larger rockets first and could launch heavier payloads into space. Their probes, however, seemed to lack the sophistication that NASA was forced to adopt with their necessarily lighter launch constraints. Throughout the Soviet Moon programs, and subsequent planetary programs, their images constantly lagged behind NASA's in terms of quality and quantity, and as such are rare to find today.

Interestingly, the colour imaging experiment on the Surveyors proved to mostly be a waste of time. After much frustration in trying to determine the Moon's real colour, it was discovered that the lunar soil was so grey that any colour seen in returned images proved to be artefacts in the camera system. The Apollo missions, begun a year after the last Surveyor fell silent, confirmed the colour of the Moon as flat grey during midday, but progressing to a light tan at extreme solar angles. It was also the advent of Apollo that signalled the end of NASA's unmanned Moon exploration. Surveyor 7 was the last US robot to visit the Moon for over twenty years.

The Saturn V Moon rocket, designed by ex-Nazi officer Wernher von Braun, had yet to be fully tested.
The Lunar Module, designed to land two astronauts on the Moon's surface, was nowhere near ready. Spooked by an unmanned Soviet Zond circumlunar flight, the upcoming Apollo 8 shakedown test would instead carry the first human beings around the Moon. Former Apollo Deep Space tracker John Saxon explains: *"They had two unmanned Apollo missions, the second one was not a real success... Man had never gone more than 850 miles above the surface of the earth at that point in time. They took a hugely gutsy decision to put 3 guys on top of that rocket in which they think they solved their problems and then send them off around the Moon. It was just mind-blowing they could do that at that point. It was an incredibly dangerous mission in a lot of ways."*

In December 1968, while the US was wracked with racial unrest, the untested Apollo 8 sent astronauts Borman, Lovell and Anders free from Earth's gravity and around the dark side of the Moon. The three astronauts became the most isolated human beings in history as Mission Control anxiously waited for their radio signal to reappear from behind the Moon.

John Saxon: *"When they finally came out from behind the Moon on Apollo 8 the guy at the front end of the tracking station was trying to find out where the source was coming from and switching things madly. I was trying to find out which communications bus he had it on and send it out to Houston. I never forget the voice of Apollo man saying, 'We have data but no voice' and I could have just pressed that button and gone up to the spacecraft and say 'we're sorting it out and it's just a small problem.' We did sort it out after that but there was that horrible delay and I still feel embarrassed about it."*

Apollo 8 made it around the Moon and, circling around the limb, saw the blue marble Earth rise above the barren lunar landscape. Anders grabbed a colour camera and captured one of the most famous images of all time. The Earthrise photo likely started the environmental movement by showing a fragile Earth lifting itself above the desolation of the Moon.

Following Apollo 8, two more test runs of the hardware were finished and finally, on July 16, 1969, three men climbed into a rocket to attempt humanity's first ever landing on another world. Meanwhile, Russia's hopes for manned lunar travel had been dashed when a bolt was sucked into one engine of their massive N1 Moon rocket's first stage during its maiden launch. The rocket exploded with half the energy of a nuclear bomb, destroying itself and most of its launch pad almost instantly. The incident left America free to attempt the first manned Moon landing alone, while Russia confidently began an unmanned Moon program: *"at less cost and without risk to human life."*

(Above) The Lunar Module descends on its own rocket on its way to the surface of the Moon. The slender contact probes extending from the landing gear can clearly be seen (NASA).

'Opposite top) One of the only photographs in existence of Neil Armstrong on the Moon (NASA). (Opposite bottom) Buzz Aldrin stands beside a science station with the Lunar Module behind him (NASA). (Top left) An Apollo astronaut descends the ladder to the Moon's surface (NASA). (Top right) The spindly Lunar Module sits on the stark Lunar landscape (NASA). (Bottom left) The lack of atmosphere causes deep black shadows on the Moon (NASA). (Bottom right) Nearly all of humanity are in this Apollo photo of the Lunar Module returning to Lunar orbit. The Command Module pilot who took the picture was the only human to be absent from theis

Commander Neil Armstrong, Lunar Module pilot Buzz Aldrin and Command Module pilot Mike Collins blasted off for a three-day journey to the Moon. In an event that united humanity for one of the few times in history, Armstrong and Aldrin separated from Collins in the Command Module, and began their descent to the Moon. Also trying to land on the Moon was the Soviet Luna 15, a robotic sample return spacecraft. This represented the Soviet's last-ditch attempt at beating the US in returning lunar samples before Apollo 11. The principle driver behind the Soviet robotic efforts, Georgy Babakin, knew that if his spacecraft was able to land on the Moon and return soil before Apollo 11, then Russia could at least claim to have been the first to retrieve a piece of the Moon. Luna 15 was orbiting the Moon at the same time as Apollo 11, causing consternation at NASA — who in the cold war paranoia at the time wondered if the Soviets were trying to interfere with their astronauts. Luna 15 achieved Lunar orbit successfully and mission controllers gingerly tried to send their spacecraft to the surface on July 20. Soviet geologist Sasha Basilevsky waited anxiously by the Caspian Sea for his precious 100 grams of lunar soil to be collected before Apollo 11. Unfortunately, the capsule never came. Whether it was a programming fault or a direct result of rushing the spacecraft preparations, Luna 15 crashed on the Sea of Crisis, ending the chance of beating the US in retrieving a sample from the Moon. Officially, the mission was declared an orbital and spacecraft systems test, to make way for a future successful landing. The US were now clear to make the first successful landing and sample return, assuming they could get down safely.

Armstrong and Aldrin's descent did not go smoothly. The Lunar Module (LM) guidance computer, critical for a safe landing, was getting overloaded and failing to process commands. 1202 and 1201 alarms were being called out by the Apollo crew. *"Give us a reading on the 1202 alarm,"* Armstrong had said, as he prepared to abort. After some frantic seconds at Mission Control, then 26-year-old engineer Steve Bales made a crucial and quick decision to override the alarm and allow Apollo to continue descent. Meanwhile, with fuel running low, Armstrong realized the supposedly smooth landing area he was plunging into was instead a football field-sized crater filled with boulders the size of cars. With less than a minute of fuel remaining, he checked the descent and raced over the surface, desperately searching for a safe area. Finally, with 25 seconds fuel remaining, the LM's footpads finally came to rest on lunar soil. Armstrong reported: *"Houston, Tranquillity Base here. The Eagle has landed."*

A few hours later, one sixth of the world's population watched as the ghostly image of Neil Armstrong carefully descended the LM ladder and just over four months before Kennedy's deadline, and place his booted feet on the Moon. This iconic moment sparked passion for many future space researchers, such as the late NASA Ames researcher David Willson: *"I was inspired when watching the Gemini and the Apollo flights. I recall, at 11 years old, when Apollo 11 landed on the moon, I declared myself too sick to go to school. My mum bought the lie and I stayed at home to watch the moon walk on TV. A wonderful experience, establishing again my fabulous space nerd credentials."*

That there was live television coverage of the Moon landing was almost a fortunate afterthought and was very nearly not included in the Apollo mission. Deep Space tracker John Saxon explains: *"The television was a recent addition. There was a lot of opposition to having TV at all on that first mission to the moon... the TV was not a high priority believe it or not. The high priority was the backpack data, maintaining communications between the backpacks and the LM on the surface and back to the earth. NASA was particularly interested in the biomedical information, the astronaut respiration, heartbeat and the spacesuit data was also going back to Houston."*

Buzz Aldrin joined Neil Armstrong on the surface, and for the next two hours and 25 minutes they trialled moving in one-sixth gravity, grabbed lunar samples and set up scientific experiments. US president Nixon also made the longest distance phone call in history and talked to the two astronauts on the Moon. Finally, having used less than half their backpack oxygen and water supplies, the astronauts returned to the LM and lifted off to join Mike Collins circling above in the Command Module. Besides being the most isolated human in history, Collins was busy photographing the Moon from orbit to help identify future landing sites. He eagerly welcomed his dirty lunar comrades, and three days later they safely returned to Earth to enjoy quarantine and a hero's welcome.

Public interest in the Moon program dropped off sharply after Apollo 11. This was unfortunate, as, although a success, Apollo 11's goals were very conservative: get to the moon, land, take off again and return to Earth. The Tranquillity landing site was deliberately chosen to be flat, safe and boring. Planetary researcher James Head explains: *"...There was this constant pressure, on the part of systems engineering, actually, where do we want to land? Well, you just can't say. The engineers were all trying to find areas that were safe to land, which was their job. But that didn't mean that it was the best place to land; it just meant that they felt that you could safely land there."*

The remaining six Apollo landing missions pushed the boundaries of lunar exploration with increasingly sophisticated equipment. Five months after Armstrong's first step, Apollo 12 made a pinpoint landing to visit a now silent Surveyor 3 probe. However, the next Apollo almost didn't make it home. Half way to the Moon, the supposedly routine Apollo 13 flight had just finished a video broadcast that none of the major television networks bothered to air, when suddenly an explosion rocked the spacecraft and Jim Lovell announced: *"Houston we've had a problem here."* As power and oxygen leaked out of the Service Module, the three Apollo astronauts began working desperately to save their lives while Mission Control frantically attempted to invent off-the-cuff procedures to get the crew back to Earth.

Communications between Earth and Apollo were also a problem. John Saxon: *"... The problem with [Apollo] 13 as far as communications were concerned was that for the first time they had put a beacon on the 3rd stage of the Saturn V rocket which was on its way to the Moon ahead of the LM... and they were on the same frequency... They never anticipated having both transponders, the LM and the S4B on the same frequency. So what had to be done was that we had to ask Houston to get the crew to turn off the LM transmitter just at the time they were having all this chaos going on with the lifeboat, trying to get the Command Module buttoned up and all the rest of it."*

Eventually, after combating power loss and carbon dioxide build-up, the half-frozen Apollo 13 crew splashed down in the Pacific Ocean. The ordeal would afterwards be described as NASA's finest hour.

A hastily reconfigured Apollo 14 LM landed on the Moon in 1971 and planned to investigate the rim of an ancient crater to try and find bedrock that could reveal the Moon's history. Alan Shepard, the first American in space, along with Ed Mitchell, desperately tried to find their planned destination but the stark landscape was making navigation almost impossible.

From the Moon Mitchell's confusion was apparent: *"Okay, I think we're very close to it. I think this crater we just went by is probably it, but it's very hard to tell... I don't see anything else that might be it, unless it's the next crater up."* Running short of oxygen, the commander Al Shepard called it: *"I don't think we'll have time to go up there to Cone Crater."* Mission Control agreed. The lunar explorers had to give up and settled for collecting samples from where they were. Turning back, it was later found they came within 30 metres of their goal.

As later Apollo missions were cancelled by the US government, the last men on the Moon pushed the boundaries of science as never before. Apollo 15 was packed with upgraded spacesuits that would allow the astronauts to explore for hours at a time and, for the first time, a car. A startled Jim Irwin exclaimed: *"Contact, Bam!"* as their super-heavy LM virtually pounded into the lunar surface. Soon after, Jim Irwin and David Scott drove their electric Lunar Roving Vehicle (LRV) to Hadley Rille, a snaking valley cutting into the Moon's floor. As late as 1967, some scientists thought these valleys were carved by ancient water flowing on the Moon.

Scott very carefully parked the LRV beside the Rille to make sure it didn't fall in it and started investigating the true nature of the valley. Photographs and returned samples showed Hadley Rille to be carved not by water but by underground lava. Over time, the top of the lava tube had collapsed, leaving a valley that looked like an earthly riverbed. Liquid water also featured elsewhere in Apollo 15. Deep Space tracker John Saxon: *"When Apollo 15 was on the Lunar surface... they sprung a leak... They had water around in the Lunar module which they had to keep mopping up... Their Lunar Module was called Endeavour. Would you believe that it was 200 years to the day that Captain Cook's ship Endeavour hit a reef up in Cape York and had to be repaired?"*

Apollo 16 landed on Descartes, part of the lunar highlands, in April 1972. John Young, who would

Working hard to stay upright on a steep hill, an Apollo astronaut samples the Lunar regolith. The Lunar Roving Vehicle is behind him (NASA).

go on to fly the first Space Shuttle, and Charles Duke spent three days exploring the highlands and brought back samples showing that impact cratering, and not volcanism, had shaped this area of the lunar surface. Apollo was showing the advantage of sending geology-trained humans, rather than robotic spacecraft, to other worlds. In three days, Apollo 16 performed more science than the long-lived Opportunity Mars rover would conduct in over a decade of operations.

As 1972 drew to a close, so did the Moon program. Following a spectacular night launch, Apollo 17 sent the last human beings ever to leave Earth orbit on the most ambitious landing mission to date. As Commander Gene Cernan stepped onto the Moon, he said: *"Houston, as I step off at the surface at Taurus-Littrow, I'd like to dedicate the first step of Apollo 17 to all those who made it possible."* The Apollo 17 site had been chosen from orbital images taken from earlier Apollo missions. One of these had shown craters that looked like they might have been volcanic vents. If true, then some of the Moon may have experienced volcanism, as the Victorian era astronomers had originally thought. Commander Gene Cernan and Harrison Schmitt, the only scientist ever to visit another world, spent three days exploring the valley up close in their LRV. In the last hours of Apollo, with their oxygen running low, Schmitt called out: *"we have orange soil!"* Commander Gene Cernan stopped what he was doing and looked over: *"Hey, it is! I can see it from here."*

Geologists in the back room of Mission Control were excited. Could this be the evidence of volcanic activity they were looking for? Looking at the astronauts' dwindling oxygen supplies, Jim Lovell, Apollo 13

(Above) An artist's view of one of the Apollo launch pads, now abandoned, as it looks today. The crescent Moon shines through the platform, beckoning future explorers.

30

Commander and part of the Apollo 17 support team, called out a warning: *"They have to leave at some point regardless of what they've got."* After hastily grabbing samples of the soil, which seemed to be volcanic deposits ejected from nearby Shorty Crater, Schmitt and Cernan left the Moon for the last time. Cernan: *"... as we leave the Moon at Taurus-Littrow, we leave as we come and, God willing, as we shall return, with peace and hope for all mankind."*

The orange soil initially caused much excitement, as planetary scientists contemplated a Moon that may have been volcanically active to recent times. This soon turned into disappointment when subsequent dating showed the volcanic glass to be billions of years old, indicating the Moon had been geologically dead for a long time.

Scientists had cause for much celebration though. The Apollo program had returned over 400 kilograms of lunar samples, which would be investigated by laboratories around the world for decades to come. Humanity's understanding of the Moon had increased exponentially and although seen as the pinnacle of the Space Race, Apollo had laid the foundations for technological innovation that has inspired the digital age of global positioning systems, environmental satellites and handheld computers. Former Deep Space tracker Clive Bloomfield wrote: *"They say technology and social change is more rapid during war. The Cold War was the catalyst for rapid change, and we can see that looking back over the past 50 years. When you think today a scientific calculator that you can buy for under $10 in a chemist has more power than all the main-frame computers in Houston during the launch of Apollo spacecraft."*

Apollo 17 marked the last time any person has left low Earth orbit, and more than one mission controller breathed a secret sigh of relief that nothing worse than what struck Apollo 13 had happened on the program. Reflecting on his time supporting Apollo, John Saxon said: *"It's all about budget. The good thing about Apollo was that there was almost no limit on the budget. You didn't build fancy stations on the ground; it all went into flight hardware. It was a huge budget as you know and there was a massive outcry as to the amount of money that was being spent on Apollo. Nevertheless, I think we were all sad when it was over."*

Running in conjunction with the Soviet N1 rocket development was a project lead by chief designer Babakin to design and fly an automated sample return mission. The project, initially designated Ye-8-5, had gotten off to a shaky start in 1968 as it faced immediate opposition from the manned Moon project team. They saw the Ye-8-5 as a hindrance to their manned program, stealing away precious boosters and flight time that could otherwise be used to qualify their space hardware. This unfriendly competition between manned and unmanned space programs has continued to plague space exploration to this day, not only with the former Soviet Union but also with NASA.

Development of the Ye-8-5 progressed anyway, leading to the construction of a 3.6 metre high two-stage monster of a craft. After entering orbit, the Ye-8-5 would descend to the ground using a main engine, then two smaller verniers. A remotely commanded drilling rig would then extract a corkscrew sample of the Moon and place it in a spherical capsule in its upper stage, the only item that would return to Earth. The lower stage would continue to function as a fixed Moon station, returning images and scientific data until its power ran out. Following the failure of the N1 rocket and abandonment of any plan to send a cosmonaut to the Moon, the Soviets did two things: Firstly, the Communist Party spread the mantra that Russia was never in the 'Moon race' and its only missions had been 'much safer' unmanned ones; secondly, the Ye-8-5 program, now being the only show in town, was rushed forward.

Twelve months later, following yet another mission lost to launch failure, Luna 16 soft landed on Mare Fecunditatis. It had landed during the lunar night and its on-board spotlights had failed to work, so the images sent back to Earth were barely discernible. Despite this, controllers managed to lower Luna 16's drill rig to the ground and extract 100 grams of lunar soil which was then deposited and hermetically sealed inside the upper stage's return capsule. Finally, 26.5 hours after arriving on the Moon, Luna 16's upper stage fired, sending its precious cargo back towards Earth. The lower stage continued to operate, sending temperature and radiation data back to Earth. Three days later, a ten-metre square pin-striped parachute opened near Dzhezkazgan (now Jezkazgan in Kazahkstan) and a helicopter homed in on the capsule's radio beacon. The Soviets finally had their piece of the Moon.

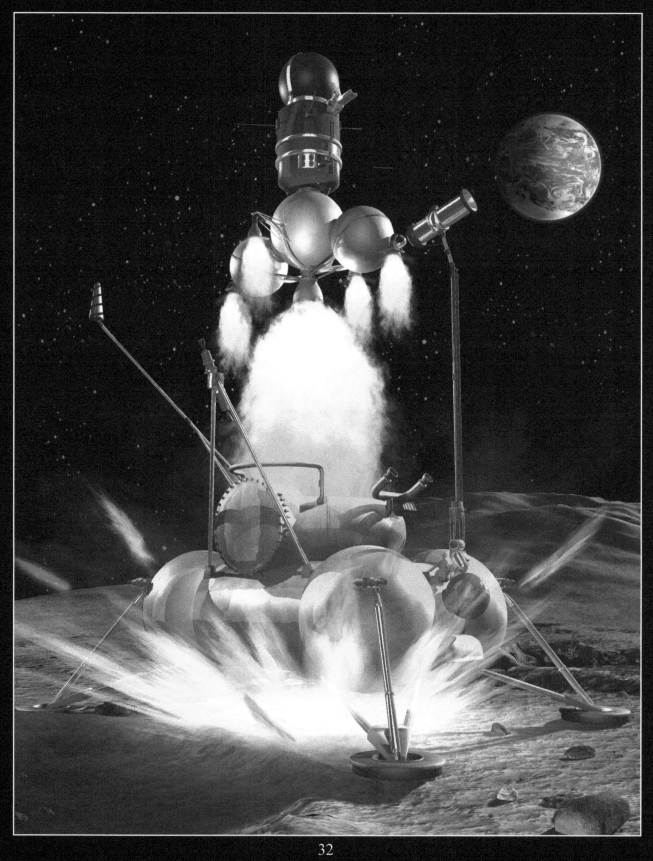

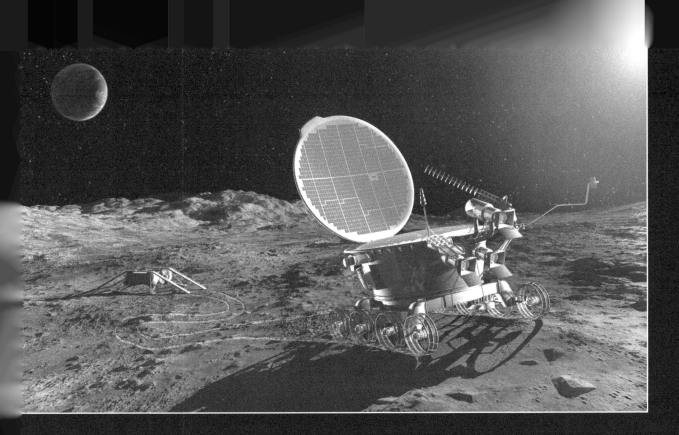

(Opposite) Carrying a few grams of precious samples, the Soviet Luna 16 blasts off from the Moon. (Top) The remote controlled Lunokhod rover departs its descent stage and begins exploring the Moon. (Bottom) The Moon passes before the face of the Earth in this deep space image (NASA).

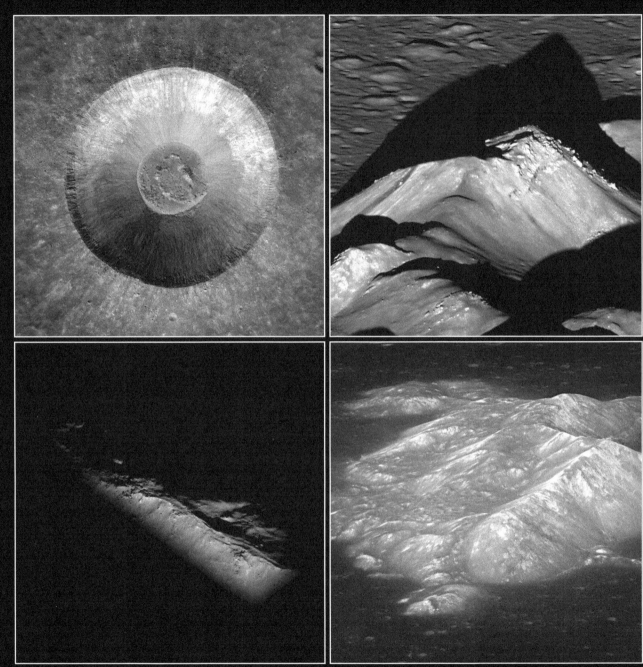

LRO has given a new look to our Moon. (Top left) An almost perfectly circular crater (NASA). (Top right) The Lunar mountains regain some of their pre-space age cragginess in this LRO oblique shot (NASA). (Bottom left) The Moon's high ground catches the rising sun's rays (NASA). (Bottom right) The mountains of the Moon rise above the surrounding lava plains (NASA). (Opposite) The full Earth rises above the Moon. South America, Africa and parts of the Middle East are visible (NASA).

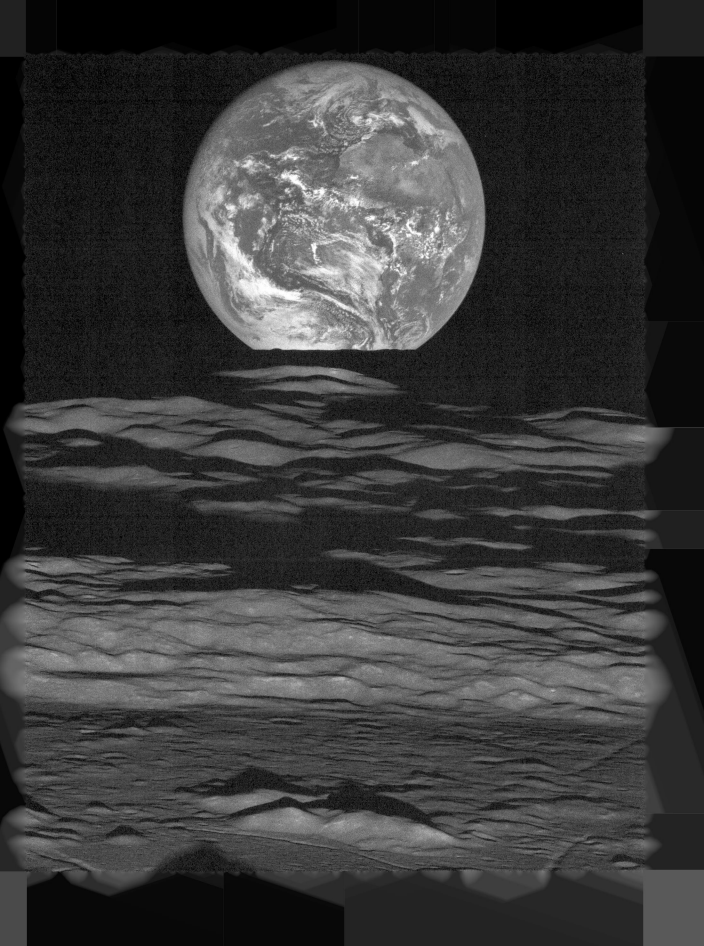

Luna 17 was next to fly to the Moon, in November 1970. The spacecraft used a similar lower stage to Luna 16 but, instead of an upper stage, it carried a rover. Lunokhod 1 was originally designed as part of the Soviet manned Moon landing project. The rover was initially designed to reach the Moon prior to a manned lander and traverse the terrain until a suitable site for a human mission was found. The rover would then park and act as a beacon to guide the incoming cosmonauts to a safe landing. Now Lunokhod 1 was a mission in itself, landing on Mare Imbrium on 17 November 1970, alone.

The eight-wheeled Lunokhod 1 was carefully driven down its landing ramp and rolled onto the lunar surface. The bathtub-shaped rover possessed a number of pioneering designs that would find themselves on rovers sent to Mars decades later. The bulk of the rover's body was made up of a nuclear-heated warm electronics box that would prevent the rover's systems from failing during the two-week long lunar night. Power was derived from solar cells on the rover's top as well as in a flip-top lid. The rover's science package carried television cameras, soil probes, and a French-built reflector that could be used to help determine exact Earth-Moon distance. Lunokhod 1 was operated remotely from Earth, though it couldn't quite be driven 'live'. A command sent from Earth took two seconds to reach the Moon. Lunokhod's operator was forced to send a command to move the rover a small distance, and then had to wait a few seconds for the images to return, confirming that the rover had moved the prescribed distance and hadn't run into any obstacles. It wasn't a perfect system, and Lunokhod 1 would occasionally sink to its axels in the loose lunar dust. Despite these occasional mishaps, Lunokhod 1 was a resounding success. Stopping only to hibernate at night, the rover was in operation almost three times longer than initially designed, returning 20,000 images and travelling over 10 kilometres.

On 21 February 1972, as the Apollo program began to wind to a close, the Soviet Luna 20 landed on the Apollonius Highlands and began drilling operations. Five days later, 30 grams of lunar samples were added to the Soviet's modest collection. In December that year, Gene Cernan shut the hatch of the Lunar Module for the last time and left the Moon, ending NASA's manned Moon program. NASA scientists turned their attention to poring over the mountains of data returned from the Apollo missions, as well as the nearly 400 kilograms of rocks the astronauts had collected during their stay. As far as the general public was concerned, exploration of the Moon had ended.

The former Soviets continued to send unmanned missions to the Moon throughout the early 1970's but, perhaps unfairly, they lacked the media exposure the West had enjoyed. In 1973, another Lunokhod was deposited onto the Moon's surface. Heavier and more sophisticated than its predecessor, Lunokhod 2 made a panorama of its landing site in Le Monnier crater before rolling down its ramp and charging its batteries. During its four-month life, Lunokhod 2 traversed hilly terrain and valleys. Two more Soviet lunar missions were successfully completed following Lunokhod 2's demise. In 1974, Luna 22 entered lunar orbit and studied the surface with a battery of imaging cameras, magnetometers and other instruments to measure the environment around the Moon. Luna 22 adventurously manoeuvered a number of times, skimming just 25 kilometres above the lunar surface at the expense of shortening its mission time. Running out of fuel, the Luna 22 mission ended barely six months after it had begun.

Two years later, the Mare Crisium received a robotic visitor, Luna 24. The robot drilled into the regolith and its upper stage soon flew off to deliver 170 grams of rock and dust to eager Soviet scientists. Soon afterward, the lower stage ran out of power and grew cold, signifying the death of the last artificial object to reach the lunar surface for decades. By 1976, interest in the Moon had already waned. Apollo was a fading memory and the 'red planet' Mars had just received two of the most complex orbiter/lander spacecraft ever to be launched into space: Viking 1 and 2. Russia was also concentrating its efforts on Venus by this time, and the Moon—so near and so far away—was forgotten.

Pick up any book about space from the 1960's and 1970's, flip to the 'future of space travel' section and fantastic predictions about manned Moon bases from the mid '90's will jump out at you. Lunar prospecting using advanced automated rovers and solar powered soil processing units would begin in the 1980's to prepare for humanity's triumphant return to the Moon a few years later; this time to stay. Unfortunately, this dream of lunar conquest has still yet to happen.

The early 1990's saw Moon exploration given a hand by the same strange bedfellow that had heralded the space race – military supremacy. During the 1980's, US President Reagan had drawn plans for a massive satellite missile defence system that would shoot down enemy warheads using high-powered laser beams. The program was aptly named 'Star Wars' and part of its justification was the unannounced Soviet space missions in the late '60's, feared to be a precursor for testing nuclear weapons in space.

Massive budget blowouts and the end of the Cold War in the early '90's killed off the Star Wars program, though the US was still keen on developing a missile defence system. In February 1994, the Ballistic Missile Defense Organization and NASA joined forces to launch the spacecraft named Clementine to lunar orbit. Clementine carried 25 scientific and technical instruments and was designed to test new sensors in the rigours of space so that they could be applied to missile defence systems. Clementine was also the flagship of NASA's new 'faster, cheaper, better' methodology, where smaller, more cost-effective spacecraft would be pushed out into space at a fraction of the cost of the 'Rolls Royce' missions of the 1970's. The little Moon orbiter took only 22 months to be built and throughout its life returned 1.8 million images of the Moon. Clementine was so successful that US President Clinton stated Clementine "constituted a major revolution in spacecraft management and design."

Some may think that the Clementine mission was merely going over old ground that the Apollo missions had already covered. However, unlike the Apollo missions, Clementine flew over the Moon's poles and discovered the possibility of a substance not thought to exist on the Moon – water ice. Deep within a crater forever hidden from the sun on the Lunar South pole was estimated to be a small lake's worth of frozen water that could one day be used to generate fuel and oxygen for future manned missions. NASA believed the ice originated from comets impacting the Moon and was preserved in the permanently frozen and dark environment of craters on the South Pole that never received any sunlight.

The lure of ice helped pave the way for a further NASA mission to the Moon, Lunar Prospector, launched on 6 January 1998. Lunar Prospector was only a small craft—1.5 metres tall—and it had no camera. In fact, the spacecraft was a giant leap backward, built cheaply and lacking a complex on-board computer. Lunar Prospector was designed simply to follow the ice. If Clementine's ice was confirmed, then interest in the Moon might be rekindled with visions of water-fuelled polar bases.

Over the course of its mission, Lunar Prospector indicated that, hidden away in the dark recesses of the polar craters, thousands of Olympic pools worth of water ice was on the Moon. This discovery was initially carefully redacted by NASA scientists until they were quite sure of the results. Then the existence of water on the Moon was announced to the world as being one of the most important discoveries since Neil Armstrong first walked on its dusty surface. Lunar Prospector appeared to confirm Clementine's results and visions of semi-permanent bases living off the Lunar ice soon fuelled renewed interest in Earth's nearest neighbour.

In 1999, Lunar Prospector was reprogrammed to change orbit, its mission completed. On board the spacecraft was 28 grams of the late Eugene Shoemaker, a legendary geologist who had studied the Moon in detail. Part of his cremated remains, as well as the rest of the spacecraft, smashed into the Moon on 31 July 1999, a few days past the 30th anniversary of Apollo 11.

Although the US had placed men on the Moon with Apollo, unforntunately,the attitude was that of 'been there, done that' regarding future exploration. NASA reasoned that Apollo had explored the Moon so well that there was little justification for sending robotic probes there, while the planets such as Mars remained largely unexplored. In the meantime, other countries, which were developing their own space capabilities, found the Moon to be quite near, and thus an ideal target to aim for. In the early 2000's, three Asian countries: Japan, India and China, all sent their first ever missions to the Moon. In regard to Japan, this nation had been designing a mission for the Moon from the early 1990's and these plans eventually evolved into a more sophisticated mission that would bring high definition television to Earth's neighbour.

Meanwhile, in India, a little-known Earth monitoring satellite group were working largely unnoticed to build Asia's first deep space craft. Starting from humble beginnings, in the 1980's, the Indian Space Research Organisation (ISRO) had developed its own launch capability and remote sensing satellites,

(These pages) Achieving the first soft Moon landing in four decades, China's "Jade Rabbit" explored the Moon
nearly 20 months, returning high definition video of its landing site (Chinese Academy of Sciences/ China
National Space Application (The Science and Application Centre for Moon and Deep Space Exploration)

almost from scratch. In 2001, ISRO felt confident enough to plan and design a mission that would fly to the Moon. Although many saw this as having no practical benefits, ISRO member Dr Kasturirangan wanted to *"... demonstrate India's capability in terms of reaching the Moon; conducting experiments in its vicinity and contribute to the overall understanding of the evolution of [the] Moon."*

The Chandrayaan-1 orbiter was announced by the Indian Prime Minister on their Independence Day, and launched in October 2008. From the time of Apollo and despite Luna Prospector's results, many presumed most of the Moon to be desiccated; none of the returned Apollo samples showed so much as a drop of water within their rocky matrix. As Chandrayaan-1 passed over the lunar pole in its orbit, its sensitive instruments confirmed the presence of the life-giving liquid, not just in the shaded craters where other missions had hinted, but in sunlit areas as well. A sub-probe was released in November, and deliberately crashed into Shackleton crater at the lunar south pole. As the sub-probe vaporised itself and its surrounds like a high explosive bomb, the debris plume was studied in detail by the orbiter. The results were telling – there was water on the Moon. This major discovery from a new player in space exploration, as well as additional evidence of water traces in other areas of the Moon, had profound impacts for future manned exploration. The presence of water opened up the possibility of creating rocket fuel and oxygen from the lunar surface, making it much easier to establish a base on the Moon.

After less than a year into its planned two-year mission, Chandrayaan-1 began to die. Thermal shielding issues were causing the spacecraft to overheat. ISRO controllers tried shifting the spacecraft into a different orbit to buy some time, but on 28 August 2009 it was all over. After completing 95 per cent of its intended mission and being the first to discover water on the Moon, Chandrayaan-1 fell silent.

In the decades since Apollo, China had been steadily increasing its space capability. China quietly built its space infrastructure, first building rockets to send satellites into space, followed by taikonauts – the nation's equivalent to astronauts.

On 24 October 2007, China's first spacecraft to orbit the Moon, Chang'e 1, was launched. This was followed three years later by Chang'e 2, a second orbiter. Through these missions, China was gaining experience operating in deep space and soon began planning for something no nation had achieved for nearly 40 years – landing on the Moon. Chang'e 3 was a both a bold step and a major jump in complexity. Unlike an orbiter, Chang'e 3 would not only have to reach the Moon but somehow reduce its massive orbital speed to a complete stop, somewhere safe enough on the surface to deploy a six-wheeled rover.

After years of design and testing, Chang'e 3 launched on 1 December 2013 on what was Asia's first ever attempt to land on the Moon. Fourteen days later, the ambitious project began its perilous descent to the surface. Like the Apollo missions before it, Chang'e 3 visually identified safe landing areas during descent. However, unlike Apollo, a trained astronaut did not perform this task. Instead, the most sophisticated electronic brain ever sent to the Moon used artificial intelligence to process electronic images and decide autonomously where to set down. 100 metres above the surface, the lander suddenly stopped descending and, using precious fuel, hovered above the ground. Image matching algorithms went to work, sifting through digital images and adjusting the lander's course to safer ground.

"It seems that Chang'e has decided on its landing site," a Chinese media commentator said. After another 70 metres descent, the lander stopped and hovered again for another critical image check. It was now or never. There was no way for Chang'e 3 to climb back to orbit; if its on-board computer failed to find a safe area now, China's hopes for its first Moon landing would be over. Descent continued, and the lander's descent rockets began spraying lunar dust in all directions, rendering the surface invisible. Finally, 12 minutes after arriving at the Moon, Mission Control in China received a signal that caused a round of spontaneous applause – Chang'e 3 had made it!

A few hours after the lunar dust cleared, the *"Jade Rabbit"* rover, strongly resembling the NASA Mars Exploration Rovers sent to the red planet some years before, began carefully deploying its solar panels and preparing to move. Explosive bolts fired, unlocking its wheels and, ever so carefully, Jade Rabbit rolled down the lander ramp. It placed two, four, then six wheels onto the lunar dust, making *"one giant leap for China,"* according to a Hong Kong newspaper.

Over the next month, Jade Rabbit explored the Sinus Iridium using instruments specifically designed to look for minerals that could potentially be mined for future lunar settlements. During this period, it discovered something that Apollo 17 had tried to find decades before: evidence of recent volcanism. *"Our analysis indicates that this young lunar mare region has unique compositional characteristics, and represents a new type of mare basalt that has not been sampled by previous Apollo and Luna missions and lunar meteorite collections,"* wrote Zongcheng Ling and his colleagues in an article for the journal Nature Communications. *"The CE-3 landing site and... analyses of the rocks and soils derived from the fresh crater near the landing site provide... ground truth for some of the youngest volcanism on the Moon."*

The mission was designed to rove for distances rivalling those achieved by the Apollo astronauts in their LRV. However, on the second bitterly cold lunar night, tragedy struck. Signals from Jade Rabbit ceased. Fearing the worst, Chinese mission controllers desperately tried to bring their sick rabbit back to life. Nothing was heard, and the mission was declared dead. Over the next month, as attention moved away from Jade Rabbit, China spontaneously began receiving signals from its rabbit on the Moon.

"It came back to life! At least it is alive and so it is possible we could save it," said lunar spokesman Pei Zhaoyu, from the Xinhua news agency. *"The temperature on the Moon is considerably lower than our previous estimation... certain components may be suffering from 'frostbite',"* said Chinese Society for Space Research deputy Jianyu Wang, following the Jade Rabbit's revival. The rover, though no longer able to move, spent the next two years returning useful science from its fixed position. By the time the last transmission was received, Jade Rabbit had become the longest-operating Lunar rover in history. China was firmly part of the new space race and, bolstered by the success of its little rover, began making plans for an ambitious landing on the lunar far side. If successful, this would be the first far side landing of any nation and would underpin China's supremacy in recent lunar exploration.

In 2009, NASA, almost playing catch-up following the Asian successes, decided to launch two spacecraft to the Moon simultaneously. The first was Lunar Crater Observation and Sensing Satellite (LCROSS), which was barely more than a collection of electronics boxes surrounding its own rocket fairing. Its simple job reflected its simple design, as stated by NASA, which was to *"... confirm the presence or absence of water ice in a shadowed crater near a lunar pole."* LCROSS would follow its rocket booster to bomb the Moon. According to NASA, *"The Centaur rocket will strike first, transforming 2,200 kilograms of mass and 10 billion joules of kinetic energy into a blinding flash of heat and light. Researchers expect the impact to throw up a plume of debris as high as 10 kilometres... Close behind, the LCROSS mothership will photograph the collision for NASA TV and then fly right through the debris plume. On-board spectrometers will analyse the sunlit plume for signs of water... and assorted organic molecules."*

On 9 October, the booster rocket, followed four minutes later by LCROSS itself, ploughed into the Moon. Mission control reported: *"Impact! Stations report [loss of signal]."* The "blinding flash" NASA had hoped for instead became a hard-to-see dust plume half hidden in shadow. This was very disappointing to amateur astronomers, who had been hyped up on social media to be prepared for a spectacular light show. As if to compensate, an explosion of legal controversy blew up after the impact. Citing the UN Outer Space Treaty, some concerned issue groups accused NASA of committing a hostile act of aggression and a violent intrusion against the Moon. While NASA made assurances that no explosives had been used in LCROSS, and that the Moon was regularly pounded by meteorites anyway, scientists managed to find what they were looking for. *"We are ecstatic,"* said Anthony Colaprete, LCROSS project scientist. *"Multiple lines of evidence show water was present in both the high angle vapor plume and the ejecta curtain created by the LCROSS Centaur impact."* As one NASA commentator mentioned: *"The argument that the moon is a dry, desolate place no longer holds water."*

Riding the same rocket as LCROSS was the Lunar Reconnaissance Orbiter (LRO). Like its sister craft, LCROSS, LRO was built quickly. In 2004, then US President George Bush had announced a crewed return to the Moon in time for the 50th anniversary of Apollo 11. Among its seven instruments was the highest resolution camera ever sent to the Moon, the Lunar Reconnaissance Orbiter Camera (LROC). As the spacecraft swooped low over the Moon, the LROC would pick out details as small as a dinner plate. As LRO

(Above) Post-Apollo Moon exploration may be more adventurous. Future astronauts negotiate boulders in a rough part of the Moon in this telephoto view (David A. Hardy).

(Above) An ideal future of the Moon would be to terraform it to become more like Earth. Such a long term project would allow water to flow on our closest world. The new Moon would need constant maintenance to prevent it from reverting to a lifeless world.

scientist Mark Robinson said: *"... you could look at human scales on the surface to find safe and engaging landing sites."*

The spacecraft was built in just four years and, four weeks shy of Apollo 11's 40th anniversary, LRO was in orbit around the Moon. One of its first tasks was to not look forward to new discoveries, but to look back at a piece of history. A camera had finally been sent to the Moon that could directly image hardware left behind from Apollo, decades before. Mission specialists scurried with orbital dynamics and instrument shakedowns to meet the 20 July deadline.

As LRO flew over the Apollo sites, the science team quickly received and processed the images for release to the public. NASA Goddard lunar geologist Noah Petro was on hand to view the results: *"We have some of the most stunning images that I've ever seen. When I first took a look at these images my jaw dropped to the ground. When you've seen something that you've never seen before in a quality that you've never seen before, it made me speechless."* Clearly visible in the imagery were the descent stages of the Apollo LMs casting long shadows on the lunar surface. Here and there, the scuff marks of Apollo astronaut footmarks, lying exactly where they had left them a generation ago. *"We can retrace the astronauts' steps with greater clarity to see where they took lunar samples,"* said Noah Petro. Harrison Schmitt, the only scientist to ever walk on the Moon, reflected on the LRO view of his Apollo 17 landing site: *"The LRO program now has provided a much, much higher resolution suite of photographs for future astronauts. Every new environment in which a geologist works is usually very different from the last but if you have learned things from your previous experience they do in fact maximize the value of your new experience."*

LRO continues to orbit the Moon today, discovering gullies and caves, and also impact craters that have

formed since the spacecraft arrived in orbit. Among the gigabytes of data are images of the six Apollo landing sites. Clearly visible are the darkened astronaut and rover trails, discarded hardware and the LEM landing stages. These images represented the first time that Apollo hardware had ever been directly photographed since the Moon landings. Standing silently on the surface decade after decade, the spent landing stages represent a legacy and a challenge to future generations. Other LRO discoveries have kept researchers busy, though of all the returned data, the Apollo landing sites seemed to have had the most impact on our recent understanding of the Moon. *"These images remind us of our fantastic Apollo history and beckon us to continue to move*

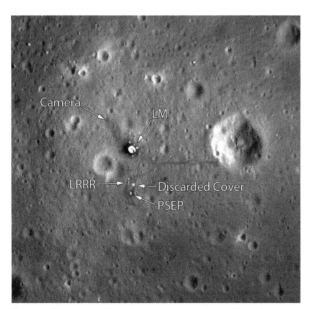

(Above) The Apollo 11 landing site as seen by LRO. Items left behind by the astronauts are labeled (NASA).

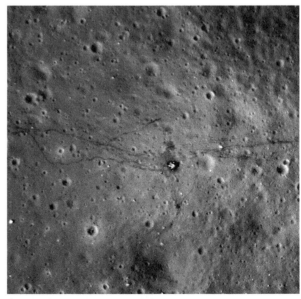

(Above) An LRO shot of the last time humanity landed on another world. The Apollo 17 descent stage and astronaut footmarks are visible (NASA).

forward in exploration of our solar system," said Jim Green, Director of the Planetary Science Division at NASA.

During the 40th anniversary of Apollo 11, some of the veteran space trackers who ensured signals from the Apollo astronauts made it back to Earth were celebrating. One of them said: *"Enjoy the 40th anniversary as there will probably not be a 50th."* As we approach half a century since Apollo 11, his words are recalled rather ominously. Both the first and last person ever to land on the Moon are no longer with us. As of the time of writing four of the original 12 who landed on another world remain, and soon no one will be left alive to tell future generations what it was like to walk on the Moon. As with a Mars landing, a return to the Moon has always been a few years away, and just out of reach.

Private companies are also trying to shoot for the Moon. The Google Lunar XPRIZE, announced on 13 September 2007, offers US$ 20 million for a team to land on the lunar surface, drive 500 metres and return high definition video and imagery. Google intended the prize to stimulate a private innovative Moon race, though a decade later the list of over a dozen candidates dwindled to five, then three, contenders. None of the competing teams were able to secure a firm launch date, and the Lunar XPRIZE may expire without ever being awarded.

In 2017, US President Donald Trump announced that NASA would again send humans to the Moon. Unlike the Bush announcement, in 2004, much of the hardware and rockets have been built and tested. Better spacesuits, allowing greater flexibility and mobility, have been designed, as well as the Orion capsule and launch system built from legacy Space Shuttle hardware. In recent years, China has pulled ahead of the US in having actually landed something on the Moon, while also testing a sample return mission to launch by the end of the decade. Only the future will tell whether Apollo was a first stepping stone for humanity's journey beyond Earth into the cosmos, or a brief period where our grandparents achieved the impossible and touched another world. Either way, the leftover Apollo hardware on the Moon continues to remind us of what we once were, or perhaps might be again.

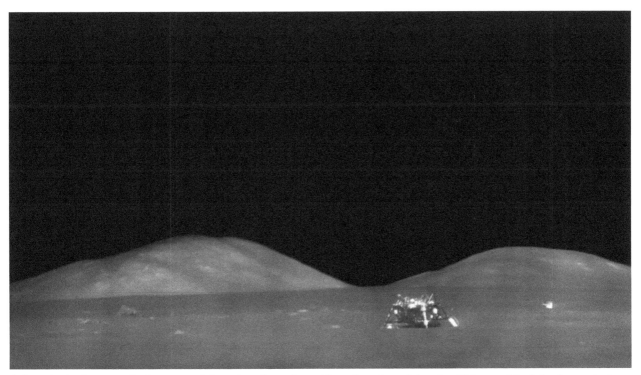

(Above) The last Apollo images of the Moon's surface. The abandoned Apollo 17 Lunar Module descent stage rests alone in the Lunar wilderness. (NASA).

(Above) A pre space-age view of a Lunar landing using designs from the British Interplanetary Society. Craggy Lunar peaks rise in the background (David A. Hardy).

The Moon's closeness to Earth has inspired humanity from prehistoric times. The first realistic drawing of the Moon was believed to be drawn by Jan van Eyck in the 1400's, predating Leonardo da Vinci, one of the world's most successful early scientists. Pre-Galileo, artwork of the Moon depicted it as a bright celestial sphere, sending its soft rays to an Earthly scene. Galileo's discoveries of the Moon and the planets as natural worlds and not crystalline spheres, permeated society. As they did, the Moon as an astronomical feature began appearing in art. In 1871, James Nasmyth used plaster models to create closeup moonscapes for the ground-breaking book he co-authored with James Carpenter, 'The Moon: Considered as a Planet, a World, and a Satellite.'

The acute angles of the sunlight and shadows near the lunar terminator misled many astronomers, including Nasmyth and Carpenter, into believing the Moon's mountains were steep, craggy features. Many of the artistic masters, including Chesley Bonestell, painted perilously steep mountains and valleys that rivalled the Himalayas, towering over winged moonships and astronauts. Lucien Rudaux was an exception to the rule. Realising the lunar mountains would be eroded by meteorite impacts, he painted them a more realistic rounded shape. Later, any disappointment of Apollo photographs showing lunar landscapes differently than expected was eclipsed by the exciting reality of people actually walking on the Moon.

For centuries, humanity has imagined many ways of reaching the Moon, from dreams, bottles of dew, or in the case of Jules Verne, a massive cannon. One of the more realistic ideas for Moon travel came from the British Interplanetary Society (BIS), for which David A. Hardy was an illustrator: *"While many plans to go to the Moon involved sleek, probably nuclear-powered, streamlined spaceships (eg. 'The Conquest of Space'*

(Above) An Apollo view of a Moon landing, showing the flimsy actuality of the Lunar Module. As shown in the photograph, the Lunar mountains are much more worn and rounded in reality (NASA).

by Chesley Bonestell and Willey Ley, and the film 'Destination Moon', both from 1950), in Britain, Ralph (R.A.) Smith, an engineer and artist with the BIS, was working with H.E. Ross on more realistic designs. The original concept from 1939 was to prove that a manned lunar expedition could be designed using existing powder rocket technology. The spaceship consisted of 2,490 such rockets, which were shed as soon as they were exhausted. Thus a kind of 'infinite' staging was used to compensate for the very low specific impulse of the motors. The hexagonal booster was 6 meters in diameter and 32 meters long, with a lift-off mass of 1,112 tons. The painting is based upon a magazine article published at the time but was actually done in 1965 at the request of the BIS.

However, in 1947 the design was changed radically due to the German liquid-fuel technology used in the V-2 rocket. It was, of course, still a step-rocket, with stages falling away as they were exhausted, and the lander was in many ways quite similar to that of the Apollo — squat, with spider-like landing legs. The post-war BIS turned to more practical near-term tasks, inclusive of holding a landmark conference in 1951 to plan the world's first orbiting satellite."

Dr. William K. Hartmann is one of the founding members of the Planetary Science Institute and has studied the Moon for decades. His work has led to the leading theory that the Moon was created by a Mars-sized body smashing into the Earth in its early history. Bill is also a leading space artist, and his art, along with his scientific work, has inspired generations of scientists and artists. In 1961 Bill was one of the few graduate students wishing to study the Moon and planets. Although being able to study under Gerard Kuiper, after

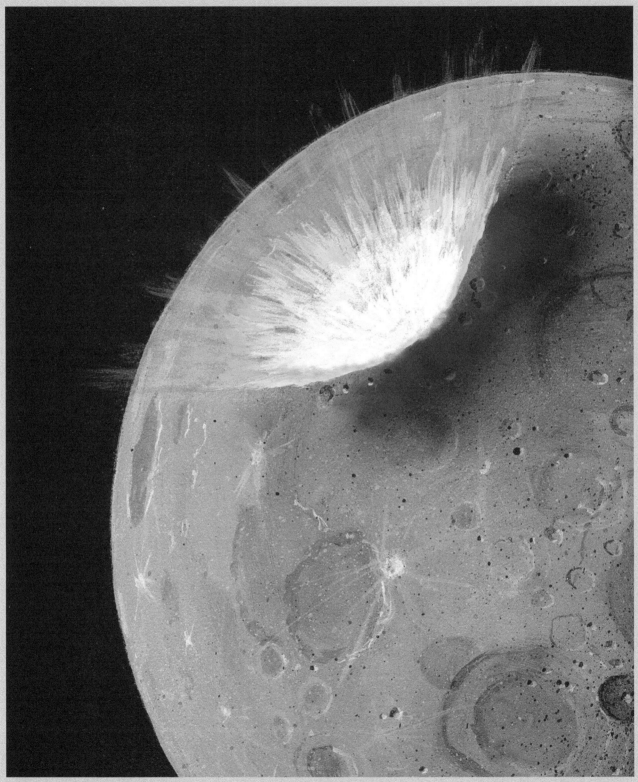

(Above) During the early period in the Solar System's history, the Moon and planets were pounded by meteorite impacts. Bill Hartmann has painted one such gigantic collision creating Orientale Basin on the Moon early in its history (Bill Hartmann).

which the Kuiper Belt would later be named, Bill was hampered by the lack of planetary science departments in universities. *"Kuiper, however, had already established a 'Lunar and Planetary Lab' and had started an ambitious program of creating atlases of the moon and a catalog of lunar craters, all limited to the front side of the moon - since the Moon keeps one side toward Earth. A Russian lunar probe had sent back fuzzy photos of the lunar far side, but the features around the edge of the Moon were poorly known because they were very fore-shortened, as seen from Earth."*

Bill was assigned to one of the atlas projects - the 'rectified' lunar atlas: *"Kuiper's idea was to project the sharpest telescopic images of the moon onto a white hemisphere, set up in a lab facility, and then moving a camera around to photograph lunar regions 'from overhead.' One day a photo had been set up that showed the east 'edge' of the moon with exceptional clarity, revealing portions of circular arcs of mountain rings. To some observers these might have seemed unremarkable, but I had been reading the scientific literature focusing on arguments about whether the craters and large circular basins were caused by volcanism or asteroidal impacts, and the circular rim structures and indications of radial fractures immediately suggested impact. I made a number of photos of this and similar images under different lighting and took them to Kuiper."*

Kuiper agreed with Bill on the importance of his finding, and graciously let the second year student co-author a publication on the discovery of the multi-ringed feature. Later age estimates showed the newly-discovered Orientale Basin to probably be the youngest of a number of giant, multi-ring basins on the Moon.

"This experience influenced me greatly in terms of the relation of art and science. In the 1960's,

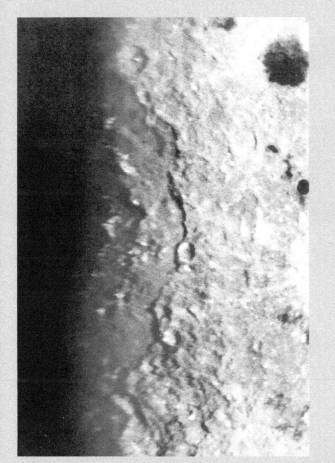

(Above left) The rectified image Bill Hartmann used to discover the Orientale Basin hiding on the edge of the Moon. The basin appears flattened when viewed from Earth (Bill Hartmann). (Above right) A painting from Don Davis of the basin, at a similar viewing angle as Bill's discovery image (Don Davis).

virtually all lunar scientists were trying to 'zero in' on the tiniest lunar details, trying to analyze details relevant to lunar landings. To me, it was stunning that we could 'back off' from the moon and see thousand-kilometer features that had not been seen before. This relates very much to the experience of a painter. Every serious painting of a solar system object involves moments of 'scientific recognition'. Many features in that painting may be well known - but then, as artists trying to depict a human experience, we realize, many other features are not yet known! Do we really know color of the Martian sky at sunset as seen from a given location on Mars? Do we really know the textures of the ice fields of Enceladus? Do we know what a night time scene on a comet nucleus would look like, with the comet tail in the sky?"

(Opposite) The Lunar Module separates from the Command Module in a NASA promotional painting. The Moon is shown as brown in colour in this painting (NASA). (Top) The Saturn 5, still the largest rocket to ever have been built, is floodlit before launch (Doug Forrest). (Bottom) Gene Cernan, so far the last human to walk on the Moon, gives a thumbs-up on return to Earth (Doug Forrest).

MARS

To the ancients, it was an angry red star that brightened every two years to declare war and conflict on those mortals below. To modern scientists, it is the most Earth-like planet in the known universe and a possible abode for life outside our own world. Of all the planets in the solar system, none has captured the human imagination, or as much attention, as Mars. For millennia, humanity has been dreaming of fantastic beings on the red planet and of travelling there to see them. More spacecraft have been sent to Mars than any other planet, to the point that Mars has been continuously peopled by ground dwelling and orbital robots for over 20 years.

So why has Mars received all, this attention? There are a variety of reasons; many practical, some not so much. On a practical note, Mars is one of only half of the planets in our solar system with a surface that can actually be landed on. Mars is also the second closest planet to Earth, being a mere 55 million kilometres from Earth at closest approach. This equates to about a six-month journey for current spacecraft, meaning that researchers don't have to wait very long between launching a Mars mission and receiving results. This is in stark contrast to missions to the outer solar system, where years may pass in waiting for a spacecraft to arrive at its destination.

Another reason is that Mars is currently the most Earth-like planet that we know of in the observed universe. Mars shares many features with our world, such as polar ice caps, volcanoes, sand dunes and a host of features that have been carved by liquid water. These latter features continue to excite planetary scientists who compare Mars with an earlier Earth. Some have used the similarities between the two planets to suggest Mars might have been habitable for life at some point in the past. Mars' proximity to Earth, its accessible surface and promise for life have made it an attractive target for artists, telescopic astronomers and space agencies.

Even during pre-telescopic times, the red star that shone brighter in skies every 26 months

attracted attention. Mars' blood-red colour was attributed to violence and conflict and, as such, the Romans named Mars after their god of war. For thousands of years, the presence of Mars was thought to dictate the rise and fall of power and the fortunes of war. In 1877, England was involved in a bloody conflict with Russia. At this time, astronomer Richard A. Proctor, who would later draw an early map of Mars, wrote: *"In the earlier part of the Crimean War, Mars shone in our midnight skies... if Mars were in truth the planet of war... if his influence poured from near at hand upon the nations of the earth... it might well be feared that the closing months of 1877 would bring desolation on many fair terrestrial fields."*

Interest over Mars wasn't just limited to war – its occasional backward or retrograde motion across the night sky caused consternation to early astronomers. This contrary motion went against the classical thinking that Earth was the centre of the universe, and all celestial objects, including the Sun, Moon and stars, rotated around it on crystalline spheres. The Martian motion, and proper functioning of the solar system, would later be revolutionised with the 16th Century discovery that all planets, including Earth, rotated around the Sun. Mars would appear to move backward as the Earth caught up with it in its faster orbit around the Sun. This motion was also seen by Australian indigenous peoples as spirits going walkabout, stopping and changing direction.

Mars is small, only half the size of Earth, and appeared as a fuzzy red disc in the great astronomer Galileo's telescope. The ability to resolve actual features on its surface had to wait almost 100 years, when Christiaan Huygens, with better instruments than Galileo, discovered a dark patch later called Syrtis Major. Later observations by Huygens and others indicated that Mars possessed ice-rich polar caps, and that its period of rotation was around 24 hours, very similar to Earth's. This was an era where society was moving

(Facing page) The Soviet Mars 3 lander separates from its mother craft above Mars as the planet experienced the largest dust storm yet observed. Overcoming the odds, Mars 3 made the world's first successful landing on the Red Planet.

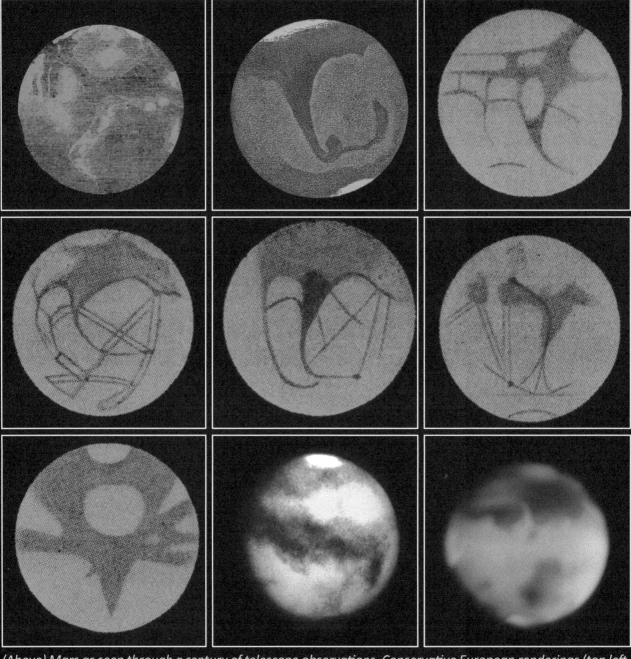

(Above) Mars as seen through a century of telescope observations. Conservative European renderings (top left, top middle) contrasted with bolder (and wrong) versions with canals (top right, middle row), and a sketch by the Australian astronomer for whom Gale Crater was named (bottom left). Even the best telescopic views just before the space age (bottom middle, bottom right) failed to solve the mysteries of Mars (Public Domain).

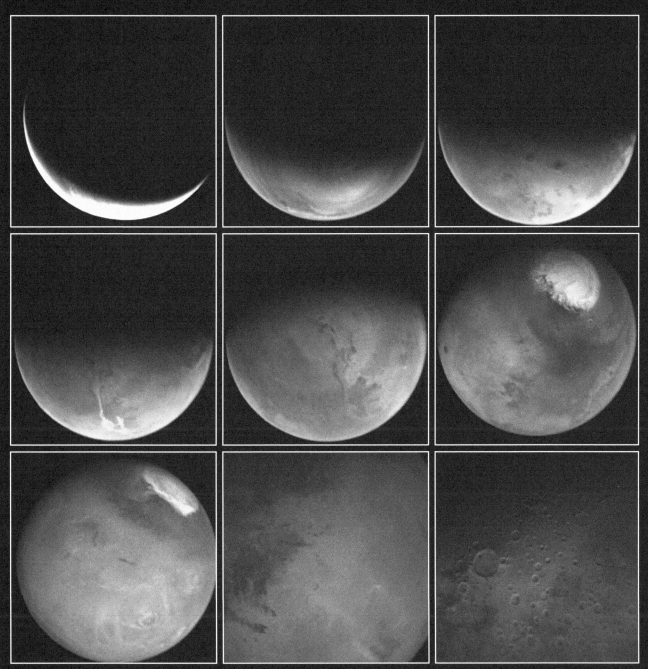

(Above) Even the low resolution Mars Express "Mars Webcam" gives us views telescope-bound astronomers could only dream about. The absence of Earth's blurry atmosphere (and spacecraft proximity to Mars) allows clear views of the Martian polar caps, volcanos and gigantic Valles Marineris (ESA).

away from the notion that Mars was a mysterious god that governed the lives of people on Earth, but was instead being discovered as an equally mysterious world in its own right. During this period, astronomers questioned whether Mars, sharing similar rotation and polar caps to Earth, might also be teeming with life. What form might this life take? There was no way to know for sure, but Mars remained a popular target for astronomers.

In 1877, an Italian astronomer named Giovanni Schiaparelli thought he saw thin, parallel lines on Mars and named these 'canali', meaning channels in Italian. This seemingly innocuous naming changed Mars exploration forever, as a wealthy Bostonian textile businessman became fed up with the academia of the day and used part of his fortune to build an observatory in Flagstaff, Arizona. Dr Percival Lowell, peering through his own telescope near the turn of last century, also called these lines 'canals' and suggested they were artificial in origin. He wrote of his conclusions in Mars and its Canals: *"That Mars is inhabited by beings of some sort or other we may consider as certain as it is uncertain what those beings may be. The theory of the existence of intelligent life on Mars may be likened to the atomic theory in chemistry in that in both we are led to the belief in units which we are alike unable to define.*

...Apart from the general fact of intelligence implied by the geometric character of their constructions, is the evidence as to its degree afforded by the cosmopolitan extent of the action. Girdling their globe and stretching from pole, to pole, the Martian canal system not only embraces their whole world, but is an organized entity. Each canal joins another, which in turn connects with a third, and so on over the entire surface of the planet."

The light and dark patches on the disc of Mars, swimming in the eternal turbulent atmosphere of Earth, fuelled speculations of a planet familiar to our own. In the late 1800's, Irish astronomer Robert S. Ball, who would deliver over 2,500 astronomy lectures to laymen audiences, wrote: *"It seems hard to decline the suggestion that the marks on the planet may really correspond to the divisions of land and water on that globe. There are circumstances which strongly suggest that water may also be present."*

Dr Percival Lowell went one step further: *"In their color, blue-green, the dark areas exactly typify the distant look of our own forests; whereas we are not at all sure that seas would. From color alone we are more justified in deeming them vegetal than marine... We conclude, therefore, that the blue-green areas of Mars are not seas, but areas of vegetation."*

Dozens of newspaper and magazine articles, as well as a host of science fiction and fantasy writers, populated Mars with fantastic aliens that stretched the limits of human imagination. The concept of a Martian civilization struggling to survive against a dying world was thought to be able to counterbalance the selfish expansionism of humanity on Earth. Conversely, such an advanced race could look hungrily at the plentiful Earth with eyes of conquest — a story H. G. Wells presented with War of the Worlds. More than one apprehensive gaze was cast to the red planet to ensure such an old civilisation didn't decide Earth was a better home and come visiting.

Many in Europe took a more conservative view of Mars. Just prior to Schiaparelli's "discovery" of the Martian canals, Richard A. Proctor had written: *"... the unchanging colour of the land regions implies they are naked and sterile...we must accept the conclusion that the land surface is arid desert"*. In 1895, Robert S. Ball, lecturing school students about the Martian atmosphere, remarked: *"As to the composition of this atmosphere we know nothing. For anything we can tell it might be a gas so poisonous that a single inspiration might be fatal to us; or if it contained oxygen in a much larger portion than our air does, it might be fatal from the mere excitement to our circulation which an over-supply of stimulant could produce. I do not think it the least likely that our existence could be supported on Mars, even if we could get there. We also require certain conditions of climate, which would probably be all totally different from those we should find on Mars."*

Wallace, co-evolutionist with Charles Darwin, was more critical of Lowell's theories: *"Not only is Mars not inhabited by intelligent beings, but it is uninhabitable"*. Such commentary was treated with scorn by Lowell, particularly when Alfred Russel Wallace had almost certainly never studied Mars through a telescope. For the next twenty years, a 'Mars fever' gripped the educated population and the planet was studied as never before. In fact, a radio broadcast of H. G. Wells' War of the Worlds in 1938 caused panic in some quarters when it was mistaken for a real news bulletin.

Throughout the 20th Century, as astronomical instruments and methods became more sophisticated, conservative views of a drier, colder Mars took hold. Independent attempts to confirm Martian canals seen by Lowell was drawing increased criticism, such as this from the highly reputed observer of Mars, Eugene Antoniadi, in Europe: *"Nobody has ever seen a true canal on Mars, and thus the more or less rectilinear canals of Schiaparelli, single or double, do not exist either as canals or as geometrical lines; but they have a basis in reality, since all are situated either on a continuously spotted irregular track, a rugged grey border, or an isolated and complex patch. Indeed, the whole of Mars presents this infinitely irregular structure."*

Jon Clarke, president of Mars Society Australia, said: *"Prior to Mariner 4, the first spacecraft sent to Mars, what was correctly known about Mars amounted to a page or two. This would have included basic facts like the planet's diameter, albedo, orbit, rotation, inclination, density, gravity, and little else. Similar basic details about the moons of Mars were also known. We thought we knew a lot more. We thought that we knew of systematic seasonal changes in colour and albedo, things like the "wave of darkening" and "blue clearing" were confidently described, if not understood. The dark areas were thought to represent significant physiographic features, even if their nature was debated. The atmosphere was thought to contain a significant about of inert gas such as nitrogen. The polar caps were thought to be frost. These known "facts" were all wrong. Then there were features such as canals and surface vegetation that many (not all) thought of as real that proved to be non-existent. Furthermore, discussion of these ideas actually hampered understanding of Mars."*

The dark patches on Mars, long believed to be evidence of vegetation, came under more intense scrutiny. The development of more sensitive spectrometry equipment enabled scientific analysis of these 'green patches' to be conducted for the first time. French-American astronomer Gerard de Vaucouleurs used the latest lead fulminate technology and painstakingly raked his new spectrometer over one of the dark patches: *"The nature of the dark areas is still unknown. It is only reasonably certain that they cannot contain chlorophyll plants similar to the higher forms of vegetal life on Earth."*

In the US, the legacy of Lowell cast a long shadow, with his treatise on Martian canals influencing generations of Mars investigators. Both Goddard, inventor of the liquid fuelled rocket, and Dr Wernher von Braun, father of America's space program, were inspired by the red planet. Lowell's vision also spurred NASA to build Mariner 4, the first spacecraft to successfully visit Mars. Former JPL Chief Scientist Dan McCleese was in charge of many Mars missions between 1994 and 2006 and commented on the influence of Lowell: *"... Percival Lowell and his intent to describe what it is that he was seeing in his telescope in Arizona just took hold of the popular press. He talked about the possibility that there might be intelligent life on the planet and that persisted for such a long time that to me it was such a great story. The people who thought there were canals said, 'alright perhaps the beings on the surface are trying to move water from the polar caps which we could see from our telescopes down to the dry equatorial regions'. This idea was present even as we sent the first Mariner spacecraft there. In fact, I have in my office a globe painted by an JPL astronomer and it shows the linear lines, it shows greens and darker colours, and these were all laid down on a sphere just as the spacecraft was travelling to the planet. It was really an incredible contrast with what the Mariner spacecraft actually saw."*

One successful space mission actually jeopardised future space exploration. This was the case for Mariner 4, later called the 'Great Disappointment', which almost killed Mars exploration overnight. In 1964, US Mariner 2 — the first successful interplanetary flyby — was just two years old, Russia's first attempts had failed, and it was America's turn to reveal Mars. Mariners 3 and 4 had a 0.02 megapixel digital camera mounted outside the bus and would reveal the true Mars to the world. Although Mariner 3 never left Earth orbit, Mariner 4 launched successfully on 28 November 1964 and headed off for Mars.

In the months preceding the encounter, Mariner 4 was tracked by NASA's Deep Space Network stations in California, South Africa and Australia. It was from Australia that a disturbing message was received by the Mariner tracking team during transit: *"Sorry, we can't track the spacecraft... galahs have eaten the radome"*.

Despite this minor issue, Mariner 4 arrived on schedule at Mars on 15 July 1965. Such was the

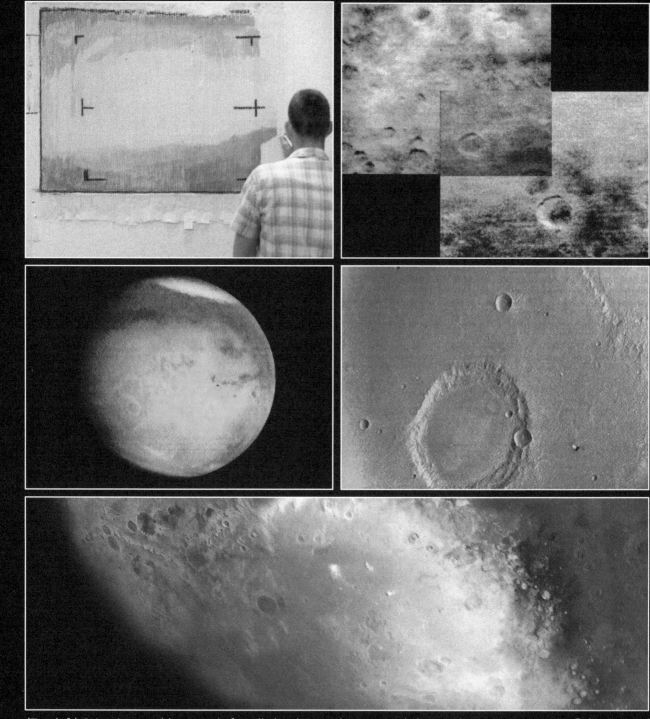

(Top left) Scientists could not wait for all the data to be returned during the Mariner 4 Mars flyby, instead choosing to hand-colour images (NASA). (Top right) The Mariner 4 frames that almost killed the space age: craters on Mars (NASA). (Middle left) Mars as it appeared from approaching Mariner 6 and 7 spacecraft. Although not realized for what they were, the enormous volcano Olympus Mons and Valles Marineris canyon are just visible on Mars' disc (NASA). (Middle right) The later Mariners returned clearer views of Mars's craters (NASA). (Bottom) A Mariner 7 view of the Martian South Pole (NASA).

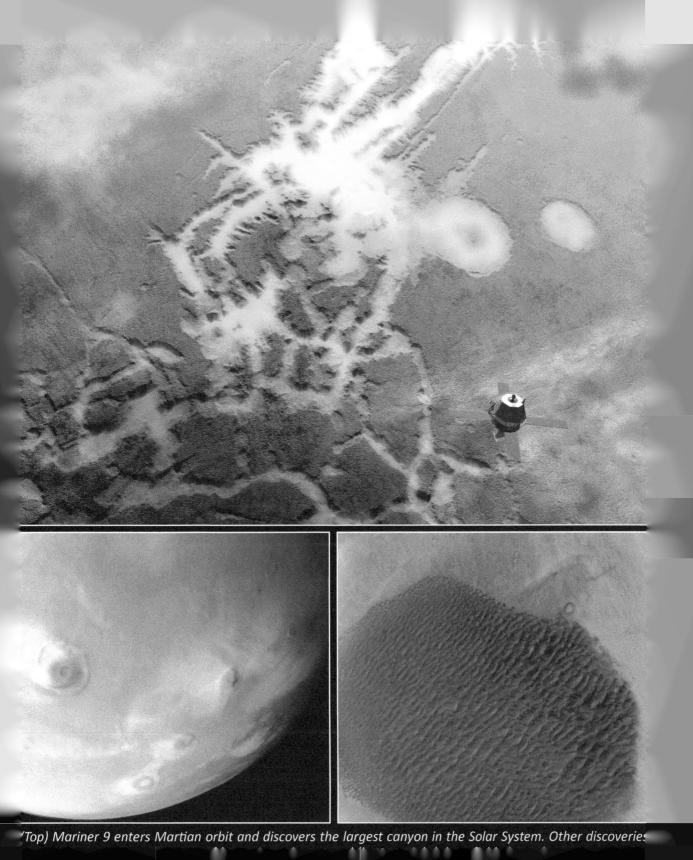

(Top) Mariner 9 enters Martian orbit and discovers the largest canyon in the Solar System. Other discoveries

excitement of the imaging team that they did not wait for the completion of the first image and set about to develop what they had. Jack James and his assistant hand coloured graph paper with crayons according to image brightness values received on a long printout. Finally completed, the first Mariner 4 image did not reveal much about Mars. The planet's hazy edge, lit by a full sun that hid most of the surface features, was all they had. Lowell's canals and primitive vegetation still seemed possible; the world would have to wait for more data.

The seventh image finally showed meaningful surface features. No sign of Martian canals or vegetation regions could be seen. In their place were craters. Active planets such as Earth have volcanoes and weather to destroy old craters. The once romantic Mars seemed to be as dead as the Moon.

Dan McCleese explained: *"There were two things that happened at the same time. One of them was that people thought it looked more like the moon and therefore quite disappointing. There weren't vast fields of organisms on the surface that might give a changing seasonal colour to the planet... At the same time of those discoveries from the spacecraft, ground-based astronomers were realizing that they had the atmosphere completely wrong. They had originally concluded from the spectra that they took that Mars had an atmosphere that was very dense like the Earth's... In fact, it's mostly CO_2 and... there was very little of it. These two things: the pictures and the knowledge that the atmosphere was very thin didn't have a lot of content of water really led people to feel a disappointment that this was probably not a living world."*

Despite US Congress' reluctance to fund more missions to Mars, the dogged determination of Mars advocates and perhaps the remaining vestiges of the Lowellian romance, led to two more Mars missions being built. Mariners 6 and 7 were far more sophisticated than their pioneering predecessors. Arriving in 1969, they flew past Mars and returned 1,100 images, covering 20 percent of the Martian surface. The higher resolution images merely showed the Martian craters in better detail; no trace of any biological, or even recent geological activity could be seen. Mars looked to be a cold, inert world, merely a larger and redder version of our Moon. In fact, Mars seemed even less interesting geologically than the Moon. Professor Robert Leighton, then in charge of the Mariner television experiments, remarked in 1970: *"Everything in the Mariner pictures indicates very gentle slopes on Mars. There are no mountain ranges, no great faults, no extensive volcanic fields, in fact no evidence of volcanic activity. You could stand in one crater on Mars and never know it – even one that appears sharp and clear in the pictures."*

In 1971, Mars would approach Earth closer than it had done since 1924. For this approach, NASA designed Mariner 8 and its twin, Mariner 9, to not just fly past a planet, but to orbit it. These latest Mariners were also equipped with colour cameras. Thanks to these developments, Mars would be seen close-up and in colour. Mariner 8 was launched on 8 May 1971, but made a short, spectacular trip into the Atlantic Ocean. Mariner 9, the remaining spacecraft, entered Martian orbit on 14 November.

A dust storm, more powerful than had ever been

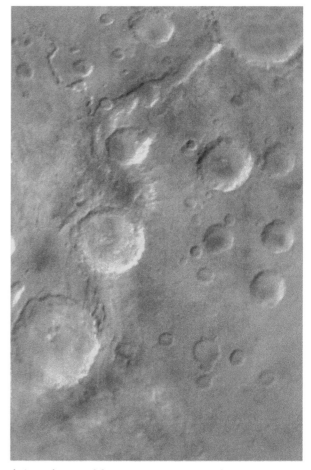

(Above) One of few images returned from the Soviet Mars orbiters. A rare colour view shows craters and a possible drainage channel (Ted Stryk).

seen on Mars before, enveloped the planet and hid the ground from view. It was this storm that almost certainly destroyed two Soviet Mars landers, arriving at Mars that year. As the storm subsided, the first features to be seen by Mariner 9 were four dark spots, thrusting up above the dust. The spots revealed themselves as the four largest volcanoes in the solar system. The largest, Olympus Mons, was so large that a theoretical Martian traveller standing on its slope would never see it completely. The planet's curvature would forever hide the rest of the volcano from view. The last of the dust storm cleared and scientists found another surprise of gigantic proportions. Cutting Mars in half lay the largest canyon system in the solar system. Seven kilometres deep in places, it could easily have swallowed up Earth's Grand Canyon several times. Massive outflow channels, braided stream beds and tributaries were photographed, suggesting Mars was once inundated with water. Although Lowell's canals were nowhere to be seen, Mars was revealing itself to be a mysterious, exotic and, at some point in history, a watery place. The infamous craters, almost the death knell of Mars exploration, were found to be restricted to the Southern Hemisphere.

Jon Clarke, Mars Society Australia President: *"The early Mars missions were as much a hindrance as a help. Mariners 4, 6, and 7 imaged one to ten percent of the planet in some detail. They showed the atmosphere was very thin and the surface had many craters. However, the areas imaged at the highest resolution missed all of the important features such as the giant volcanoes, polar caps, Valles Marineris, and outflow valleys. So, Mars was incorrectly seen as Moon-like with minor ground ice collapse and a trace of atmosphere. This mis-comprehension probably led to the abandoning of the manned Mars program and the Mars Voyager unmanned missions."*

Despite the setbacks, for the first time since 1964, Mars seemed more hospitable to life, as well as more geologically active and interesting. Mars had earned a closer look and already a mission was on the drawing board to bring back even more data and find out if life existed on Mars. The mission would be called Viking.

Viking was designed from the successful Mariner series spacecraft that had already explored Venus
and Mars. NASA proposed a conjoined spacecraft containing an orbiter that would circle Mars and a lander that would touch down on the surface. The Viking landers were built to contain experiments to monitor the Martian environment, to send back pictures and test for life. Nuclear batteries would provide the 50 watts of power required to run the lander. Problems in the initial design process cropped up from the start. Dr Tobias Owen, involved in the Viking project at the time, explained one problem: *"At that time… everyone thought the atmospheric surface pressure on Mars was around 83 mb. By 1967, ground-based spectroscopy had demonstrated clearly that the surface pressure was around 7.5 mb, too low for a parachute descent."*

There was no way a parachute would be able to slow a lander down enough to survive in Mars' tenuous air. Soft landing rockets would be required to help slow the spacecraft. The newly designed camera systems for the landers also presented issues – their power usage and weight would have to be low enough for the lander constraints yet would have to be able to survive massive temperature swings, near vacuum conditions and ultra-fine dust. Getting permission to take the cameras outside the laboratory proved surprisingly difficult, as some team leaders didn't want their expensive and irreplaceable equipment to leave their safe clean-rooms. The excursion team had wondered how they could be expected to build a camera that could survive a descent to Mars when they couldn't take it a short drive down the road on Earth.

By July 1976, after a journey of over half a billion kilometres, Viking had safely entered Martian orbit but, after weeks of searching, had found nowhere safe to land. The landing plains, selected months earlier from Mariner 9 images, proved to be full of probe-destroying geology. The celebratory July 4 touchdown was postponed, as 800 new images were pored over in an attempt to find a safer landing place.

Finally, with the Viking lander team near the end of their tether, the project manager had made a decision: The jeep-size lander, protected in a saucer-like aeroshell, departed forever from its orbiter to begin its 30-minute journey to the surface. Back in the control room, what little nervous conversation there had been died down to a cemetery-like silence. All knew this was the most hazardous part of the mission. Four Russian landing attempts had failed. Richard Bender, Chief Landing Engineer, called out the landing events as they happened: *"400,000 feet… 74,000 feet… 2,600 feet… Touchdown! We have touchdown!"*

Cheers and applause broke out as the news of a successful landing sank in. Viking 1 was down.

Finally, Viking 1 had landed and was taking the first picture within seconds of landing, in order to safeguard against any catastrophic failure. First soil, then rocks, and finally the Viking lander footpad resting safely on Mars was revealed, the whole process taking twenty minutes. The second image taken immediately afterwards was even more magical. Planned to be a full panorama, Viking revealed a surprisingly Earth-like desertscape of sand, rocks and dunes. Tim Mutch, head of the lander imaging team, couldn't contain himself: *"It's just a beautiful collection of boulders, a geologist's delight!"*

The cameras were proving to work better on Mars than they ever had on Earth. A sombre revelation would later give pause to the Viking 1 team on how lucky they had been — a three metre long, one metre high Martian boulder lay just eight metres away from Viking's camera, affectionately dubbed 'Big Joe'. Viking 1 missed becoming a billion-dollar space wreck by mere metres.

The first colour image was received at JPL on schedule. Almost immediately, a problem became apparent. No one actually knew what the true colours of Mars were supposed to be. Former deep space tracker John Saxon: *"On Viking. How do you get the colours right? They thought they had the answer. They put a colour chart that the camera could look at and they'd know. But what they didn't think of properly was that the light that was falling on the colour chart was not sunlight as we see it. But that it was going through [the Martian] atmosphere which was different."* Scientists had been expecting an almost black sky caused by the ultra-thin air, with just a faint tinge of blue near the horizon. Instead what they received was a bright reddish glow bathing both ground and sky alike. Not believing the results, and fighting an eight hour deadline, the image was hurriedly reprocessed to show a blue sky and grey rocks, a supposedly more logical scene.

The press had a field day, comparing the colours of the scene with rocky deserts on Earth. However, the blue sky lasted less than a week. Further data supported the presence of a red sky. Somewhat sheepishly, a reprocessed version of the image showing the true salmon pink Martian sky was released to the press. The imaging team was rather nonplussed when asked by a reporter whether the Martian sky would turn green in the next release. What no one had counted on were the tons of finely grained dust particles suspended in the Martian atmosphere, making the sky much brighter than anyone had predicted.

The images poured back to Earth by the hundreds, the Viking team struggling to keep up. Dr Tobias Owen, involved in the Viking project, described a typical day in the life of the Viking team during this time: *"Long hours at JPL, supplemented by homework to analyze the data as they came in, and plan for next series of experiments, listen to reports from other teams, endless discussions of all these exciting data! Now and then a press conference [was held] when we tried to convey both our excitement and the new knowledge we were acquiring."*

Viking 2 entered Martian orbit eight weeks after its twin and as with its predecessor, an exhaustive search was undertaken to find a safe landing site. On 3 September, after some heart-stopping moments when Viking's scheduled landing time came and went, flight controllers detected Viking 2's signal and the control room exploded in shouts and cheers. It, like the Viking 1 lander, returned multitudes of stunning Martian panoramas, and reported on the intensely cold, dry Martian environments. Soon attention turned to the Viking biology experiments to determine what, if any, life could survive on the red planet.

Squeezed into one cubic foot of space within each Viking lander were three experiments that would normally fill a terrestrial laboratory room. The Pyrolytic Release (PR), Gas Exchange (GX) and Labelled Release (LR) experiments would moisten, feed, cook and fry Martian soil in the hope of detecting microbial activity. Following this analysis, a final, fourth experiment would try to detect any organic compounds to clinch the presence of life on Mars.

The results of two of the tests would later become so contentious that debate over the them would rage for

(Facing page top) Missing a mission-ending boulder by mere metres, Viking 1 descends to a touchdown on Chryse Planitia in 1976.

(Facing page bottom) Viking samples the weather of an alien sky. At sunset the salmon sky will turn an eerie blue. Both Viking landers lasted long enough to track seasons over Martian years (NASA).

decades, and lead to contraversy surrounding the LR experiment designer, Dr Gilbert Levin. Levin had developed the LR experiment during his days as a public health engineer, as a way of testing water for human consumption. Eight days following Viking 1's successful landing, the command was given for Viking's soil scoop to dig a sample of Martian soil and deliver it to the biology instrument for processing. The first tests delivered promising results though the researchers were cautious, citing the possibility of exotic chemical reactions, as opposed to the presence of living microbes. It was Levin's LR experiment that would capture the media's attention: *"At Viking's two landing sites, 4,000 miles apart, my [LR] experiment, got results that satisfied the criteria for life established before the mission. I and my indispensible co-experimenter, Dr Patricia A. Straat, were very excited, but we were cautioned to await the results of the Viking Gas Chromatograph-Mass Spectrometer (GCMS) that was to identify the organic matter everyone knew had to be on Mars."*

So astounding were the LR results, that a printout of the experiment data was solemnly signed by the team. The ramifications of the results were such that a press conference was held even before the results of the GCMS experiment had come in. Before long, the world was being told by the media that life on Mars was almost a certainty. Even as the press releases went out, doubts from the biology team were increasing. The presence of oxidising chemicals in the Martian soil could account for the positive reactions in all of the biology experiments run so far. Scientists looked to the GCMS to resolve the life issue once and for all. If it could detect carbon-based compounds in the Martian soil, then the case for life on Mars would be strengthened. Gil Levin explained what happened next: *"The GCMS failed (and I use this word meaningfully!) to find any organic matter. NASA and its coterie of scientists immediately opted to believe the GCMS, and said, "no organic matter, no life," and has pretty much maintained that position ever since."*

No organic compounds could be found by the GCMS in the Martian soil by either lander, even despite

(Above) Viking 2 digs trenches in an attempt to search for life on Mars. If the lander had been able to dig a little deeper, it would have found underground ice. As it was, the ambiguous (and mostly negative) search for life results set back US Martian exploration by two decades (NASA).

Viking 2 shoving aside a rock to retrieve soil protected from direct sunlight. Tobias Owen summed up the final verdict: *"Viking showed that the surface materials on Mars were even more highly oxidized than we had thought: not only was there no indication of biological activity, there was no evidence for the presence of organic compounds."*

As far as NASA was concerned, Mars was a sterile desert. Centuries of speculation of life on Mars had suddenly ended, and dissenters from this view, such as Gil Levin, were largely ignored. Dr Dan McCleese added: *"I think the consensus in the scientific community of today is that Viking did not find organics, that the surface is very, very pristine, that there are organics that you can find in the interior of rocks but that's not what's been seen by Viking. There was so much press and to some extent science hype about the possibility that it would find life that there was a huge disappointment. That disappointment changed the course of Mars exploration so that we did not have another spacecraft go to the planet for two decades. Mars was basically written off as an interesting place to visit with our robots. The "life" thing had so tainted the expectations about why you would be interested in Mars that nobody at NASA wanted to talk to us. This was in case the "L" word would come up and they'd say, 'well we know it's a dead world and we're not going to go back.' ... Really the impact of Viking was that it closed our eyes to Mars as an interesting place."*

Following the success of Viking, Mars would not be landed on again for 20 years. Throughout this

period, one of the regrets of the Viking team was not being able to move the lander once it was on the surface. To truly explore a planetary surface, the landing craft would need to be mobile, with the ability of negotiating terrain to reach areas of interest. A rover-like vehicle would be ably suited to this task.

The result of this thinking was a small roving vehicle, about the size of a microwave oven, as part

(Above) One of the last images of Mars returned by the Soviet Union. The Fobos 2 spacecraft captured this image of the Martian moon Phobos shortly before suffering a catastrophic failure. Its sister craft, Fobos 1, had already failed part-way to Mars (NASA).

(Top) Wheeled rovers on Mars. A panorama captured by Mars Exploration Rover Spirit (NASA/Don Davis). (Bottom) The cat sized Sojourner, the first ever roving vehicle on another planet, investigates a rock (NASA).

(Above) The Martian sun rises behind Spirit as it uses sophisticated sensors to investigate Mars.

(Above) Off-world driving. Mars Rover Opportunity looks back on its own tracks as a dust devil swirls across the Martian plains (NASA/Don Davis).

(Facing page top) Spirit works hard to investigate rocks that may have existed in an ancient hot spring

(Facing page bottom left) The aluminium on which the US flag is displayed originated from the Twin Towers from September 11 (NASA).

(Facing page bottom right) The world's first Martian mountain climber, Spirit negotiates the

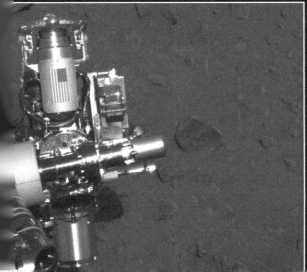

of a novel lander mission designed with a budget ten percent of that of the Viking missions that preceded it. Sojourner, as the rover was later named, was made possible through a combination of technological advancement, an engineer playing around in his garage, and by a borderline hoarder. Firstly, computers had taken off as a 'thing', meaning that the central processing unit required to control a small rover could fit in a lunchbox as opposed to a suitcase. Secondly, Don Bickler, while playing around with metal struts in his garage, devised the novel rocker bogie suspension system that would be used on all future Mars rovers. Finally, Big Blue, a monster rover left over from the Apollo days, was covertly moved between storage sheds by another engineer, saving it from disposal in case it would be "useful" one day. On the lander side, the design costs were reduced by deciding not to carefully descend via expensive rockets, but rather smack into the planet using airbags and bouncing to a soft landing.

Never before had a lander mission been designed, built and tested for such a small sum of money. Top end, custom-designed equipment was dropped for off-the-shelf solutions, while the end goals of the Pathfinder mission were pared down to the bare essentials: land on Mars in one piece, take at least one colour panoramic photograph, deploy the Sojourner rover.

Finally, after years of last minute testing and fixes, on 4 July 4 1997, the Deep Space Network station, in Spain sent a message to the mission control room: *"I see a weak signal..."* Pathfinder was on the Ares Vallis flood plain of Mars. Over the next hour and a half, Pathfinder deflated and retracted its airbags and prepared to deploy the little Sojourner rover on the Martian surface the next day. As Sojourner raised itself up and gingerly rolled off its platform until all wheels were coated red with Martian dust, the rover team realised that, for the first time in history, a mobile vehicle was operating on Mars. Sojourner helped Pathfinder capture the imagination of the public once again, with some modern help. For a generation of people born after the last Moon landing, Pathfinder was their first surface mission and there was a new tool to keep up-to-date: the internet. Mission scientist Matthew Golombek explained: *"This was one of the first global internet events, with interest peaking on Monday (when folks got back to work and their computers). A total of about 566 million Internet 'hits' were registered during the first month of the mission, with 47 million 'hits' on July 8th alone, making the Pathfinder landing by far the largest Internet event in history at the time."*

Images from the mission were being released to the World Wide Web almost immediately, allowing people at home to view the explorations of the Sojourner rover for themselves. In 1997, Pathfinder began a process of imagery dissemination to the internet that has continued for every JPL mission to this day. It was not until Princess Diana died later that year that Pathfinder's internet hit record was broken.

Although Pathfinder was designed to last 30 days and Sojourner just a week, both machines

operated for nearly three months. During that time, the lander had returned over 17,000 images and Sojourner had traversed over 100 metres of terrain, sampling a number of rocks in the process. Data received from the dynamic duo hinted that the Pathfinder landing site had once been more Earth-like and been subjected to torrential flooding by water early in Martian history. The mission had exceeded all expectations and ushered in NASA's modern plans for Mars exploration.

Pathfinder, along with Mars Global Surveyor – another successful low-cost Mars mission – promised a virtual gravy train of cheap to design and cheap to build missions that would turn Mars inside out. Unfortunately, in 1999 the gravy train derailed. Two back-to-back Mars missions failed, mainly because not enough money had been spent in testing. First, an erronous imperial to metric conversion sent Mars Climate Orbiter to burn up in the Martian atmosphere. Next, a gravity switch installed upside-down caused the ambitious and cheap Mars Polar Lander to shut off its descent engines early, splattering it so well that its remains are yet to be found.

Dan McCleese: *"... I can remember a very senior committee getting ready for the Mars Polar Lander sitting and evaluating what had gone wrong with the Mars Climate Orbiter and they said, "we are the best and brightest that the nation has, we will not fail to land." And so when it did fail everyone was just devastated and totally embarrassed."*

Two years later, the US suffered a devastating terrorist attack that shook its society to the core and overshadowed the successful insertion of Mars Odyssey into Martian orbit. The tragic loss of the Space Shuttle Columbia in 2003 led many to question whether NASA could be trusted with continued space exploration. Given these events, the pressure on Steve Squyres for his Mars Exploration Rovers (MER) Spirit and Opportunity to land safely on Mars and perform was overwhelming. As it was, MER was the successful contender of the many proposals that Squyres had submitted over 16 years, and NASA had allowed not one but two rovers to be built. Spirit and Opportunity, their names given in a student contest, were massive upgrades to Sojourner, the last

(Facing page) A Mars Exploration Rover drives into the Martian wilderness.

(Right) A Mars Express view of the Martian north pole. Two of the massive Tharsis volcanos can be seen near Mars's limb (ESA).

71

Two years late and over-budget, the mini-car sized Curiosity Rover brought instruments to Mars that the Mariner pioneers could only dream about (NASA).

rover sent to Mars. The golf cart-sized robot geologists contained all the engineering and instruments for them to leave their landing site forever and explore up to a football field worth of Mars a day.

"*The instrumentation on-board these rovers, combined with their great mobility, will offer a totally new view of Mars*", said Dr Ed Weiler, who helped administer the mission. Their design, including the cameras, promised to offer the most human-like view of Mars ever achieved, while a custom made arm complete with rock abrasion tool would provide: "*...a microscopic view inside rocks for the first time.*"

Dan McGleese: "*We had to succeed and we had to do something new and very exciting and so the rovers became really the major approach. We had the Pathfinder lander so we knew that rovers worked on the surface and that you could do science but we wanted to do something really bold and the two landed rovers were the way to go. When JPL conceived of the mission with a lot of committees and everything overseeing the process it decided that one rover was all that was necessary. But the head of NASA at the time, Dan Goldin, said you are going to send two, because I don't want to risk a failure.*"

Finally, after years of testing and last-minute fixes, both rovers landed on Mars in early 2004: "*We're two for two! One dozen wheels on the soil,*" said flight director Chris Lewicki. Both rovers found evidence of past liquid water. Opportunity literally bounced into a crater filled with sedimentary rocks and water-eroded blueberries (hematite concretions). Spirit became the first Mars mountaineer by driving three times longer than planned to distant Columbia Hills. By chance, Spirit's front wheel jammed and, digging a trench, discovered evidence of an ancient hot spring. "*Spirit's unexpected discovery of concentrated silica deposits was one of the most important findings by either rover*", said Steve Squyres. "*It showed that there were once hot springs or steam vents at the Spirit site, which could have provided favorable conditions for microbial life.*" If found on Earth, most geologists would have routinely classed these as fossilised life. However, there was no way for Spirit's limited instruments to find out for sure, and soon afterward the rover itself struck disaster and got bogged in loose sand. All efforts at freeing Spirit failed and the rover ceased to function in 2010. Her twin, Opportunity, completed the furthest extra-terrestrial traverse in history. Opportunity, at the time of writing silenced by a global dust storm, lasted so long that some of the current mission team, such as Bekah Sosland Siegfriedt, were in school when Opportunity first landed: "*I certainly never thought I'd have an opportunity to work on Opportunity,*" she said. "*That only became possible because this mission has been going so incredibly long... I'm loving that I can be a part of this team now.*"

Spirit and Opportunity also gave the US some solace after 9/11. The attack occurred just over a kilometre away from where rover components were being built in Manhattan. Rover team member Stephen Gorevan was riding his bike in lower Manhattan when a plane hit the World Trade Centre: "*... I stopped and stared for a few minutes and realized I felt totally helpless, and I left the scene and went to my office nearby, where my colleagues told me a second plane had struck. We watched the rest of the sad events of that day from the roof of our facility.*"

Although feeling helpless about the attack, the engineers decided to make a tribute to the 9/11 victims by building part of the rover arms out of metal from the World Trade Centre. Metal was delivered to the robotics company and made into abrasion shields for the rover arms. The quiet tribute to 9/11 has been a part of Mars since 2004.

Following MER, left over hardware from a previous project was able to fulfil a polar mission in 2008 as the aptly-named Phoenix. Racing against time, before the onset of the polar winter, Phoenix sampled the weather, detected snowfalls and dug up parts of the icy Martian surface. One of the most surprising finds of the mission happened directly underneath the lander. Phoenix's braking rockets had blown away the topsoil, revealing pristine ice underneath. Some of this ice had spattered on the lander's legs, and sequential images showed this ice appearing to melt and form spherical blobs. If true, this controversial find would prove the existence of liquid water on present day Mars.

Although airbag landings had been used successfully three times on Mars, something different was needed to land the next rover NASA was building. The Mars Science Laboratory, later named Curiosity, was the biggest vehicle ever to be landed on the surface of Mars. Weighing close to a tonne, the airbag

(Above) Rising from the ashes of Mars Polar Lander, Phoenix samples the ice-rich soils near the Martian north pole. Although the soil could theoretically grow asparagus if transported to Earth, whether life could exist in this environment remains an open question (NASA).

landing system would not work for this mission. A member of JPL's landing team, Adam Steltzner, said: *"Unfortunately, we don't have fabric here on Earth strong enough to build airbags that would work for a rover the size of Curiosity: the bags would shred, not giving Curiosity any protection."* Curiosity was also too heavy to carry on a beefed-up Pathfinder-style lander. Adam Steltzner, by then leading the team of landing engineers, came up with a radical solution: use a sky crane. This never-before-tried process used a rocket descent module that would lower the rover gently to the Martian surface with four nylon ropes, before flying away and crashing. *"It looks a little bit crazy. I promise you, it is the least crazy of the methods you could use to land a rover the size of Curiosity on Mars,"* said Steltzner to reporters. NASA worked hard to test for every potential problem (the sky crane descent module crashing on top of the rover was a serious consideration), delaying the launch by two years and blowing out the budget to US $2.5 billion. Finally, in August 2012, the 'seven minutes of terror' – where Curiosity would land, or crash, on the Martian surface – had begun.

Canberra locals travelled to Tidbinbilla, one of NASA's deep space tracking stations in Australia, to watch the Curiosity landing live. Mars was favourably in Australia's skies at the time for Tidbinbilla to carry the signal of Curiosity's progress. The visitor's centre was packed, all watching the images of nervous engineers at JPL Mission Control, as they monitored the progress of the spacecraft. An animation was also showing on screens at Tidbinbilla to show what was supposed to be occurring on Mars. The time lag between Earth and Mars made it impossible for mission control to intervene if anything went wrong. Although watched from a world away, the landing sequence strongly echoed that of Viking, decades before.

Mission Control were spectators as they called out critical goalposts of the Curiosity landing: *"Vehicle reports entry interface... We're beginning to feel the atmosphere as we go in here... Parachute has deployed... We're down to 90 meters per second at an altitude of 6.5 kilometers and descending... We are at altitude 1 kilometer and descending... Sky crane is starting... Touchdown confirmed! We're safe on Mars!"*

(Top and bottom) Mars shows much evidence pointing to a wetter past. Many think Aram Chaos was once

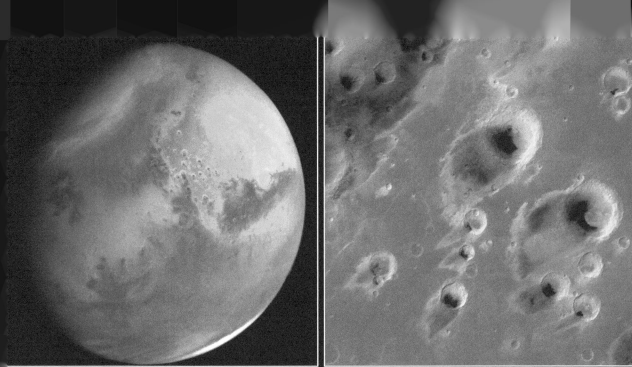

(Top) Two Martian views from the first ever Indian spacecraft to visit the Red Planet (ISRO).
(Bottom) Fog rises above Newton Basin that was once shaped by Martian rain.

(Above) Mars' closest moon Phobos skims over the landscape in this oblique shot. Some explorers think that a crewed Mars mission should land on Phobos first, using it as a way station before going to the surface. The little moon itself is slowly spiraling into Mars, and may form a ring in eons to come.

The visitor's centre at Tidbinbilla erupted into applause as hugs and cheers were exchanged at Mission Control. Curiosity was safely down and almost immediately started returning images of its landing site, Gale Crater. Named after an Australian astronomer, Gale Crater was chosen above over 50 regions of Mars as the favoured landing site for Curiosity. The rover carried an array of cameras, soil samplers, and for the first time, a Mars laser that could blast rocks up to 15 metres away. Dan McCleese: *"We have now got to the point where Curiosity is making the kind of measurements on the surface that geologists would make on the rocks in their laboratories. And that level of sophistication is absolutely amazing and is a consequence of sending ever increasingly difficult and rewarding instruments every 26 months."*

Following weeks of analysing data returned from the landing site with an abundance of cameras and instruments, NASA announced: *"Rounded pebbles that hint of an ancient river, and a delta, where a river emptied into an ancient lake. Lakes spanned the crater floor, leaving mud that built the lower part of the mountain, inch by inch. The lakes came and went, and eventually dried up. But groundwater persisted, leaving minerals behind."* Curiosity revealed Gale Crater as perhaps the best part of Mars yet explored that was most favourable to past life. The ground on which Curiosity rolled was once underwater as part of an inland lake. Streams of water knocked the corners off the rocks, rounding them, just like on Earth.

Was the Lowellian vision of a habitable early Mars finally coming true? The story of Curiosity is still unfolding and may one day come up with an answer. The nuclear powered rover is slowly being driven up Mount Sharp, a 5-kilometre high mound in the centre of Gale Crater. As Curiosity climbs, she will be moving forward in Martian history. It is a race against time, as the rover's wheels have started to fall apart. Small pinpricks, then major tears, have appeared on the wheels, due to rocks sharper than expected wearing away up to 60% of their designed life. The rover team remain optimistic: *"All six wheels have more than enough working lifespan remaining to get the vehicle to all destinations planned for the mission"*, said Curiosity Project Manager Jim Erickson. *"While not unexpected, this damage is the first sign that the left middle wheel is nearing a wheel-wear milestone."*

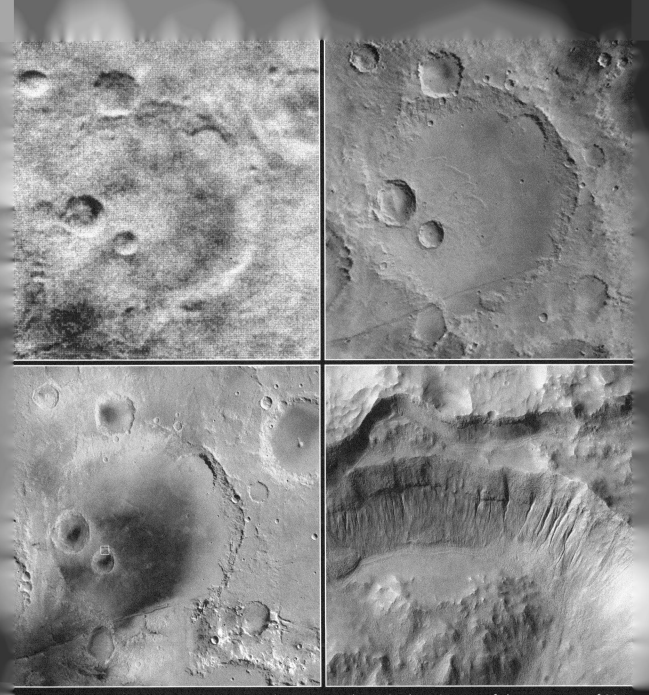

Five decades of development in Martian exploration are shown in these images of the same part of Mars (NASA). (Top left) Mariner 4 was the first to image this crater, followed by Viking (top right), Mars Express (bottom left), and finally HiRISE (bottom right). The HiRISE camera, able to image objects the size of a football...

(Above) A comet passes close to Mars and is visible after sunset over the fantastic geology o

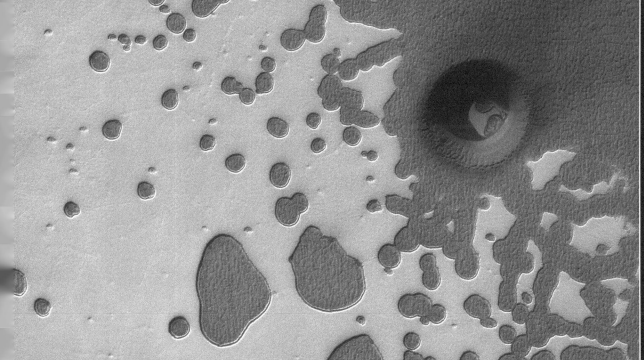

(Top) Wind is the dominant process on Mars today. Sand dunes like this one are found in many places of Mars. Like Earth, some of these dunes move over time (NASA).

(Bottom) Polar carbon dioxide ice is subliming, forming Swiss cheese-like terrain in the Martian spring (NASA).

Imaging Mars from orbit

is technically easier than sending a lander to the surface. Floating serenely above the abrasive dust storms and the lander-killing temperature shifts, the Mars orbiter has nothing but ageing electronics and dwindling attitude control gas to worry about. Following the end of Viking, a NASA team proposed sending a new orbiter to Mars without a camera. Dan McCleese: *"The first spacecraft to go back was thought of initially as a very low-priced mission. And the scientists who were managing the space exploration program at NASA felt that the most expensive thing you could do on a spacecraft is to take pictures. And if we wanted this mission to be cheap then that spacecraft which was called Mars Observer should not have a camera. And then it would be a cheaper mission. Well that idea was wrong, and eventually the NASA leadership said no way are we sending a planetary spacecraft without a camera and so while it's true there were a lot of pictures I think that trying to do it cheaply really drove the mission but that was then reversed by people who understood that the pictures were really the way of conveying the exploration to the public."*

Advances in imaging technology allowed space engineer Mike Malin to convert the 'cheap' imaging proposal to a Martian telescope. Image resolution of his instrument would rival that of aerial photography taken of Earth. In 1991, the billion-dollar Mars Observer, as the mission was later called, was lost three days before entering Martian orbit from a ruptured fuel line. However, copies of most of the instruments, including Malin's camera, were able to be flown on a budget orbiter, Mars Global Surveyor, in 1997. This later mission, and the orbiters that followed, have proven so successful that Mars is the most accurately mapped planet in the known universe (Earth's clouds and oceans make this hard to achieve for our planet). The orbiters have mapped large swathes of Mars down to 25-centimetre resolution, discovering gullies, evidence of subterranean water and surface ice activity, and revealing the fantastic geology of Mars. In addition to mapping, orbiters play a vital role in acting as relay stations between landers on the surface of Mars, and Earth.

Although the US

has dominated Mars exploration, other nations have launched missions to the red planet. So far the European Space Agency (ESA) has got it half right. Mars Express and ExoMars have returned a wealth of data since arriving in Mars orbit. Unfortunately, their landers, Beagle 2 and later, Schiaparelli, failed. Beagle 2 was the first semi-privately funded Mars mission and its loss was particularly devastating to its creator, Colin Pillinger. Perhaps even more devastatingly, later images from orbit revealed Beagle 2 had almost worked but was unable to deploy the last of its solar panels and uncover its radio antenna. Schiaparelli, in an echo to the failed Mars Polar Lander, shut down its rockets early and exploded on the Martian surface.

The old Soviet Union and the new Russia have tried for Mars almost more times than the US. All of their missions either failed or their results were eclipsed by more sophisticated US probes. One of the failures almost sparked World War III when its debris was mistaken for a Soviet nuclear attack during the Cuban Missile Crisis. The latest attempt, Fobos-Grunt, became stuck in Earth orbit in 2012, before finally burning up, months later.

In 1963, the Soviet Mars 1 suffered a mission-ending transmitter failure, but nevertheless became the first human-made object to reach Mars. The massive Mars 2 and Mars 3 actually made it to the red planet, just in time for the global dust storm that had affected Mariner 9 and returned images of a dust covered Mars from orbit. They also deployed two landers into the swirling dust, one of which was never heard from again. The Mars 3 lander, on the other hand, continued working and when tracked altitude readings levelled out, the Soviets knew that Mars 3 had made the first ever successful landing on Mars. Scientists then impatiently waited for the lander to begin transmitting the first picture of the surface of another planet. A teleprinter stirred to life, showing a noisy image. Then, after 14.5 seconds, the data stopped. The lander was never heard from again and the failure was later blamed on the dust storm overwhelming the lander's electronics. Although controversial, the image appeared to show an upside-down view of the Martian landscape with lighter sky in the background. If this was the case, then the Soviets, and not the US, were the first to image Mars from the surface. Opinion remains divided as to whether Mars 3 actually imaged the Martian surface or, as most now believe, was signal preamble.

Recently a new nation launched for Mars: India. Reversing the trend of failures before success, the Indian

Space Research Organisation (ISRO) succeeded in reaching Mars with its Mars Orbiter Mission (MOM) on its first attempt. Moreover, it succeeded in the face of international derision. Some commentators refused to believe a developing country could successfully fly a spacecraft to Mars. ISRO spectacularly proved them wrong by keeping its design simple. MOM relied heavily on existing Earth-orbiting satellite hardware, and the use of space-proven technology kept the costs down. In fact, MOM cost less to run than the Hollywood movie Gravity which was released the same year MOM launched. Much of the Western criticism was unfounded, quite offensive, and didn't even bother to address the massive morale boost MOM's success had on India's scientists and engineers.

The impact of MOM was felt across India and has inspired the next generation of engineering students. One of them said at a recent space conference: *"Long before MOM, back in 2008, India launched its first orbiter to the moon, Chandrayaan-1. This was India's first exploration mission that discovered liquid water on the Moon and it left a deep imprint on me as an undergraduate student. I still remember I used to closely follow the mission updates and read about the mission on my way to classes. It was very exciting to see that my country had decided to embark on its space exploration phase. Then in 2012, the government approved plans for a Mars mission. What really impressed me was the efficient manner in which ISRO progressed from mission concept to launch within a time period of 3 years. Mangalyaan (or Mars Orbiter Mission) was launched on the 5th of November 2013, at a time when I was working with a spaceflight team at NASA Ames, developing a bioscience payload for the International Space Station. At the time, the mission was mainly covered by the Indian media and it was only upon its Mars Orbit Insertion on 24 September 2014 (incidentally my birthday!) that the whole world took notice. For me, it was a moment that India delivered a clear message to the world that planetary science missions do not need to be overtly time and cost intensive. The sheer grit and technical rigor of the mission was clear in the achieved orbit precisions, India was among the first nations to achieve a successful Mars mission on its first attempt. Personally, I would love to be a part of the Indian Mars exploration program. The exploration roadmap presented by the ISRO chairman includes a second orbiter mission in 2020. Hopefully, I would be one day a contributor to the nation's grand Martian adventure!"*

Following half a century of space exploration, what is the real Mars? Why do we keep sending spacecraft there? Planetary researcher Dr Graziella Caprarelli offered her idea: *"With more recent missions we have better spatial resolution of the observations you can make… This has not only revealed new features that you couldn't see before but the characteristics of features that were observable before have changed dramatically from our perspective. For example, in the past we thought the northern plains were this large flat expanse... We are [now] able to observe things at different scales and we start seeing that they are not flat at all. We can see that roughness is a characteristic of the surface and it may very well reflect what is happening in the sub-surface so we do understand much, much more as long as we continue to send missions to Mars."*

Perhaps the greatest finding is that Mars is not Earth-like, nor Moon-like, but Mars-like. The Lowellian Mars, bustling with intelligent life desperately irrigating crops via canals, has gone forever. Instead, what we know of Mars today is a surprising world that is exciting and mysterious in its own right. Each new mission and discovery adds new layers of understanding and raises new questions. Like Earth, water has played a key role in shaping Mars throughout most of its history. Early Mars was a much wetter place, with oceans of water carving outflow channels reminiscent of the Lowellian canals of the early 1900's. Whether the abundant water was freely deposited on the surface via rainfall or locked under sheets of ice is a topic of robust discussion and disagreement among researchers today. Most agree that the role of liquid water has reduced greatly over the aeons, though has not entirely vanished. Smaller-scale gullying on young craters, and orbiter images of slope streaks that darken during the Martian summer, offer tantalising glimpses to a world that is not totally dry and dead.

Other processes worked on Mars as well. Massive volcanoes, faulting that has almost split the Martian surface in two, and billions of years of wind have built the Mars we know today. We now know that the simple answers we were looking for even a generation ago no longer exist. Mars is just too complex to look for simple fixes and sound bites. There is plenty left to explore on the red planet for generations to come.

Sometime in the future, after humans finally journey there, a new generation of people may be born on Mars. Unlike anyone before them, they would no longer call Earth home, knowing only rusty soil and salmon skies.

Where does that leave Mars on the scale of possibly hosting life? Has our knowledge of Mars increased the possibility of life there? Opinion is divided. Dan McCleese: *"The answer to that question is really at the root of our exploration of Mars. Curiosity has shown that the planet was habitable in the distant past, that there was water present that there were sources of energy, the sun, that there are layers of rock present on the surface that show all of the components that are required to make life possible and that the ancient climate was also conducive to making a habitable world. So, it's really opened our eyes to the possibility that in fact there may be evidence of life on the surface."*

Dr Graziella Caprarelli: *"The short answer for me is no... From the time of Schiaparelli it was thought there were people living there and that then naturally there was Mariner which went the other way, that it was not possible to have any life on Mars, there was no water, it's not active, so you had the two extremes... I would say that the jury is still out and we still do not know hence we need to keep researching if the focus of researching on Mars is about finding life past or present, but certainly we also cannot rule out 100% because we do find traces of microenvironments that on Earth are capable of sustaining life. So we still don't know."*

NASA Ames researcher the late David Willson: *"What changes have taken place since the Viking mission concluded 'no evidence for life' in 1976? I think there has been a major change in the understanding of the possible existence of life on Mars. Now it is thought that ancient life could have existed in the distant past and extant life (life that exists in recent times) could exist as well. There are many pieces of evidence. Firstly, the Viking mission mass spectrometer result of "no carbon in the soil" has been questioned by the discovery of perchlorates in the Martian soil by the Phoenix lander. The mass spectrometer heats up soil in an oven until it is an ionised gas, then passes the gas over a magnet to distribute a spectrum of molecules of ascending masses on a detector. The pyrolytic (heating) process results in the soil perchlorates reacting with organic carbon, destroying the organic molecules. Thus, Viking's mass spectrometer instrument built to detect organic compounds was actually destroying them."*

The world that probably inspired more people to become involved in space exploration than any other planet has still refused to give up its secrets on this enduring question. Perhaps the only way to find out for sure is to send humans there. Serious plans to go to Mars have existed since Apollo's conclusion, however, they have tended to be "two decades away" since the 1980's. Dan McCleese: *"I was a young scientist at JPL [in the 1980s] we had a committee of distinguished people come to the laboratory to speak to young scientists and engineers about the future... They asked, 'When did I think humans would land on the surface of Mars?' ... I said ...the earliest we could send humans was 2025. The committee went crazy. They said that they'd never heard a more pessimistic estimate of the future, particularly from a young person... Well, clearly we're at least another decade, possible two, away from seeing this happen which would make it 2035, 2045 at the earliest? So it's always kind of been a goal that's always beyond our reach since the 1980's."*

Since the 1980's, a host of new contenders, including private companies, have joined the quest for Mars. Elon Musk, PayPal founder and CEO of SpaceX, has been working on plans to send a capsule to Mars. The United Arab Emirates and China are planning Mars missions in the future. ESA, despite the Schiaparelli crash, are pressing ahead with their Mars plans. Even if Mars may not be the first thing on NASA's list, it certainly has been a priority for other nations. In the meantime, Mars shines brightly in the sky every 26 months. The 2018 opposition saw the launching of NASA's Interior Exploration using Seismic Investigations (InSight) mission, designed to investigate subsurface activity on Mars. Assuming a safe landing, InSight will be the first to analyse what lies beneath the crust of another planet. Also for the first time, briefcase sized satellites using low-cost cubesat designs have hitched a ride on InSight and signal data about the larger craft back to Earth.

Other organisations are also advocating for increased exploration of Mars. In the late 1990's university students in Colorado started Mars Underground, which has since become the Mars Society. The non-profit group has over 5000 members across 50 countries that are active in a number of Mars projects. One notable project is the Mars Desert Research Station where small crews occupy a simulated Mars habitat and perform activities expected to be conducted on Mars. Research into robots that might one day assist Martian astronauts is also a strong focus of the Mars Society.

Despite investing much of his personal fortune and countless hours of effort into studying Mars, Percival Lowell was profoundly wrong about the Red Planet. The Martian canals, irrigation streams, dying civilisations and the idea of Mars as an abode for life, led astray much of US Mars exploration for decades, making the impact of the Mariner 4 mission harder. Many might question the soundness of the preservation of the Martian view for so long, when there was mounting evidence to the contrary, even before the Space Age. Without Lowell's visions of this 'exciting Mars', space exploration as we know it might never have happened. Many of the space pioneers, including Wernher von Braun, were inspired by the fantastic Martians in the Barsoom series penned by Edgar Rice Burroughs, and works from other science-fiction writers. Would these first pioneers, and a willing public to back them, have tried so hard to reach into the unknown if they knew evidence of life on Mars would still be elusive over a century after Lowell's 'Mars and its Canals'? The search for life outside of Earth remains one of the biggest drivers for unmanned space exploration. Announcements of the existence of water or habitable conditions for life on Mars continue to make headlines, and more missions continue to be sent to the Red Planet than anywhere else in the Solar System. Perhaps, this is the greatest legacy Percival Lowell left behind – a robust and enduring interest in space exploration that seeks to answer one of the last enduring questions of humanity: Are we alone?

(Above) A Mars Society Australia project to fast-track rover technology for Mars science.

(Above) Early artwork illustrating a Martian canal (David A. Hardy).

So far, Mars remains the most Earth-like planet that can actually be studied close-up. As a result, the Red Planet has inspired more than its fair share of artwork. In fact, we have imagined our way to Mars through artwork, books, film and television more than any other place in the Solar System. Much of this early art, although wrong, inspired the space age as we know it today. One of these pioneers is David A. Hardy, who being one of the longest established space artists, lived right through the inception and advancement of the space age. The painting on the top left is one of Hardy's early visualisations of Mars before the start of the space age. David A. Hardy: *"When I became interested in astronomy, around 1950, it was generally accepted that because Mars has white polar caps, and dark areas (often thought to be greenish, probably because of contrast with the red surface) appear to spread in a 'wave of darkening' in the Martian spring, there must be vegetation. The most extreme version of this, proposed by Percival Lowell, was that there are dark lines which appear to connect the various areas. These had been observed by the Italian astronomer Schiaparelli, who called them 'canali', meaning 'channels', but Lowell interpreted this as 'canals'. This inevitably led to a belief in intelligent Martians, attempting to save their drying and dying planet by guiding water from the melting poles. We now know that the observed colour changes are due to no more than dust being moved around by the very strong winds, causing planet-wide dust storms which reveal and obscure lighter desert and darker rocks."*

Scientific understanding of the "real Mars" was certainly not smooth, to and fro-ing between a desert with sparse vegetation and a dead Moon-like world. We know a lot more about Mars today. Over the decades increasingly sophisticated robots, such as Curiosity, have revealed a fascinating and very diverse planet (top right). Yet, even with gigabytes of scientific data, Mars still continues to inspire art. In fact, some artists are

(Above) The Curiosity Rover's view of Mars (NASA).

regularly synthesizing actual spacecraft data with their creative talent in order to generate artwork firmly embedded in reality. This type of work provides a 'this is what you'd see if you were there' look. Art of this kind is very useful for designing mission concepts, movie special effects or simply trying to visualise the geology of a planet. The top artwork overleaf is an example of this latter case. The artist generated a scene of a Martian crater using orbital imagery and elevation data from the highest resolution camera ever sent from another planet. This crater is particularly interesting as over 20 gullies have cut into the far crater wall. On Earth, gullies are a sign of liquid water erosion, so the discovery of similar features on Mars was very exciting. This crater also sports a flow of ice rich material, similar to a rock glacier on Earth. This has crawled off the gullied wall, to form a series of ridges near the bottom of the crater. Was Percival Lowell not so wrong about water on Mars after all? Martian gullies, and other apparently water-carved features, remain a hot research topic for scientists studying the Red Planet.

Space-age Mars also inspires art for art's own sake. As an increasingly impatient human race looks to the Red Planet with an expectation to one day visit, in a way Mars has become part of our society. Not only that, Mars is also incredibly beautiful. From gigantic volcanoes to the almost impossibly large Valles Marineris, Mars owns a diverse set of landscapes. Space artist Eileen McKeon Butt has painted a hemisphere of Mars, floating in space (bottom image, overleaf). As Eileen said: *"Mars is a forbidding and inhospitable place, but this may not have always been so. There are signs of ancient oceans on Mars, and the red planet once had a denser atmosphere than it does today.*

We once believed that modern Mars was a totally lifeless desert, but we now know that it has periods where briny water (recurring slope lineae) runs down channels in its mountains. Now, scientists are actively on the hunt for signs of life, and young astronauts are training for what they hope will be our first manned mission to Mars! My painting is inspired by everything we're learning about Mars, and its potential as humanity's next home."

Finally, some artists use the returned spacecraft images themselves. Experienced space artist Don Davis has reprocessed imagery from a variety of missions, including the sunset view of Earth from Mars by a Mars Exploration Rover. As Don explains: *"Another fun aspect of space image processing is colorization from a black and white image. With knowledge of what one is 'aiming for', ... a result can be achieved that would compare well with what a color camera would have obtained. In the case of the Spirit rover's last landscape a detailed color panorama was gathered from many color filtered pictures. A striking black and white panorama of wider angle NAVCAM pictures near Sunset provided what to me was the most dramatic lighting of any Martian 360 degree scene ever obtained, and I used the color scene made under noon time conditions to color this carefully retouched black and white panorama, making sure every rock in the scene had its distinctive color."*

(Opposite top) An oblique view of Martian gullies. (Opposite bottom) A contemporary painting of Mars based on spacecraft data (Eileen McKeon Butt). (Below) The Earth rising over Mars (NASA/Don Davis).

CHILDREN OF SOL

Venus was among the first planets to be discovered by the ancients. So bright that it can occasionally cast shadows on a moonless night, the nearest planet to Earth took turns being a morning and evening star to our ancestors. Its brilliant radiance was also a thing of majesty and the planet was named after such goddesses of beauty. Visions of Venus fuelled the Aristotelian view of a corrupt Earth being surrounded by perfect spheres rotating around for eternity.

Smaller, faster Mercury was harder to spot, never straying far from the Sun and never in the same part of the sky for long. In pre-telescope ages, Mercury would appear as a small star in the evening sky, before vanishing and reappearing before sunrise. Eventually, the morning and evening 'stars' were discovered to be one and the same planet, and was named after the messenger of the gods. Like Venus, Mercury was seen as yet another perfect sphere circling around the Earth, along with the Sun, Venus and three other planets known at the time.

Along with the Moon, bright Venus was one of the first targets for Galileo Galilei's newly-built telescope. As he wrote in The Sidereal Messenger, Gallileo made a profound discovery: ... *I began to look at [Venus] through a telescope with great attention... At first... Venus appeared of a perfectly circular form, accurately so, and bounded by a distinct edge, but very small; this figure Venus kept until it began to approach its greatest distance from the sun, and meanwhile the apparent size of its orb kept on increasing. From that time it began to lose its roundness on the eastern side, which was turned away from the sun, and in a few days it contracted its visible portion into an exact semicircle ... After completing its passage past the sun, it will appear to us...as only sickle-shaped, turning a very thin crescent away from the sun...*"

Galileo discovered Venus had phases like the Moon. Transformation from a full-lit sphere to a thin crescent was only possible if, like Mars, the other planets, including Earth, orbited around the Sun. Galileo also reasoned that Venus and the other planets were not self-luminous, but were simply reflecting light from the Sun to be seen from Earth. Other astronomers, such as Johannes Hevelius, soon revealed Mercury to exhibit phases, though not much else was visible on the tiny disc swimming in the eyepiece of these primitive telescopes.

Venus, despite being much bigger and much closer than Mercury, was also frustrating efforts to identify features on its surface. In the 19th Century, R.S. Ball, while favourable for Venus being Earth's sister planet, explained the difficulty in conjecturing about it surface: "*Indeed, the features of Venus seem so ill-defined and vague that we cannot rely on them sufficiently to pronounce with any certainty on the interesting question as to the time Venus takes to rotate on her axis.*"

Venus was identified early on to have an atmosphere. Rare transits of the planet in front of the Sun made a teardrop appearance, as solar backscatter illuminated parts of the atmosphere to Earth-bound observers. The transits, occurring in pairs only once every few lifetimes, were highly sought-after events, and many explorers, including Captain James Cook, travelled far to witness them. A Venusian transit was recorded by Russian astronomer Mikhail Lomonosov in 1761, where Venus' outline was blurred against the sun. In the few years before German astronomer Johann Schroter's telescope was confiscated by the invading Napoleonic army in 1814, he also noted that the edges of crescent Venus extended further than they should have. The existence of an atmosphere led to much speculation as to what surface might lie underneath it.

During the 19th Century, many astronomers thought the neighbouring planets to be more exotic versions of our own. Human imagination was too limited to envisage the highly variable, poisonous worlds the Space Age has since revealed. As Venus was very similar in size and shape to Earth, thoughts of the Victorian era naturally gravitated towards a steamy version of our world. As R.A. Proctor wrote in 1870, because of its similarity to Earth: "*On the whole, the evidence we have points very strongly to Venus as the*

(Facing page) ESA's Venus Express enters the domain of Venus, closest planet to Earth. Once thought to be Earth's idyllic twin because of its size and nearness to our own planet, spacecraft such as Venus Express have since shown the Venusian surface to be more like a blast furnace.

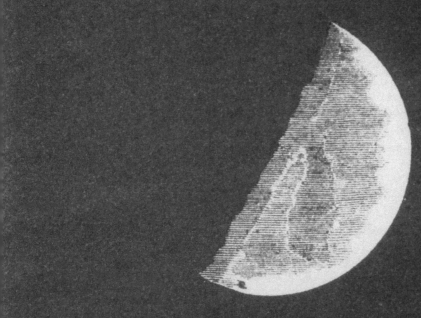

(Top and bottom) The featureless clouds of Venus frustrated pre-space-age astronomers. Although the discovery that Venus showed phases like the Moon reinforced the idea of a sun centred solar system, what lay beneath Venus' clouds could only be guessed at. (Public Domain.)

'Top and bottom) Mercury was, and remains, a harder target to resolve through a telescope. Mercury's ultra-fast orbit meant that viewing windows were fleeting. These 19th century telescope observations were restricted to showing details of Mercury's orbit and phases as seen from Earth (Public Domain)

(Above) A view of the surface of Mercury as imagined by the 19th Century astronomer R.S. Proctor. A bloated sun creates steamy conditions where an ultra-humid atmosphere is the norm.

abode of living creatures not unlike the inhabitants of earth."

In the case of Mercury, those wishing for an Earth-like world gave the planet an atmosphere, dense with clouds caused by increased heating from the Sun. In 'The Poetry of Astronomy' R.A. Proctor wrote: *"The air of Mercury is thus heavily laden with moisture seems... likely enough... No choice seems left but to adopt one or other of two general inferences respecting the possibility of life on fiery Mercury. Either we must believe the conditions under which life can exist vary much more widely than anything known here... or else... this small planet is not only at present unfit to be the abode of life, but cannot have been inhabited in any past era, and can never become habitable hereafter."*

Proctor, like other astronomers of the 19th Century, was hampered by Mercury's small size and being only visible in bright skies. The closest planet to the Sun is a difficult world to study. Ground-based observations were difficult because it rarely ventured far enough from the Sun for a good look. Indistinct markings were seen by some astronomers such as Schiaparelli, who also thought he had discovered canals on Mars. Percival Lowell, also of Mars fame, even went so far as to 'identify' canals on Mercury. These canals, he reasoned: *"although roughly linear"*, were natural in origin and didn't signify the presence of intelligent Mercurians. If Lowell had followed the same reasoning for Mars, space exploration may have turned out very differently. Other observers, such as E.M. Antoniadi, also saw light and dark markings on Mercury, though these 'discoveries' did more harm than good. Based on perceived observations of markings of Mercury, its 'day' was wrongly assumed to be the same length as its year, 88 Earth days, and the innermost planet was thought to present the same face to the Sun.

The 20th Century brought with it better telescopes and instruments to analyse the chemical compositions of our neighbouring worlds. No one could find any atmosphere around Mercury, which Proctor

(Above) Some conservative pre-space age astronomers refuse to guess on what lay beneath Venus's clouds. Persistent dark markings seen from Earth-based telescopes were thought to be peaks of tall mountains poking through the clouds.

had speculated on decades before. Transits of Mercury, more frequent than those of Venus, failed to show any of the blurring or teardrop effect seen in transits of Venus.

The nature of the Venusian atmosphere had stubbornly refused detailed analysis, and theories abounded of lush tropical life thriving beneath eternal rainclouds in an oxygen-rich atmosphere. It was not until 1932, at the height of the Great Depression, that two astronomers used sensitised film to identify carbon dioxide as a main constituent of Venus' atmosphere. To the imaginations of Fred L. Whipple and Donald H. Menzel of Harvard University, cloudy Venus became a world covered in a soda-pop sea, with limestone-crusted islands. In 1955, radical British astronomer, Sir Fred Hoyle, instead envisioned Venus as an oil-king's dream. He reasoned sunlight splitting water into its elements, then recombining to create enough oil and natural gas to support an industrial economy for millennia.

In 1956, a tool that had helped Great Britain survive its darkest hour during World War II – radar – became advanced enough to be used to investigate another planet. Venus was close enough to Earth to be able to capture at least some of the returns. As radar beams were bounced off the surface of Venus, for the first time, hard data about its environment was revealed. Firstly, a day on Venus was found to be an enormous 240 Earth days long, and backwards. If the Sun could be seen on Venus, it would rise in the west, taking days to do so, then agonisingly slowly, make its way to an equally long sunset in the east. Further analysing the radar waves, scientists found surface temperatures to be hundreds of degrees celcius. These initial findings provided a glimpse into the poisonous, hellish world that Earth's first successful interplanetary spacecraft would later reveal.

Despite its hidden surface, Venus was an attractive target for the nascent Space Age. It was the closest planet to the Earth, placing it within easy reach of the temperamental rockets available at the time. The short distance also placed trip times within the lifetimes of the newly invented electronics needed for space exploration. In 1961, just four years after launching the world's first ever satellite, the Soviets sent their first spacecraft to Venus. It was a step too far. Venera 1 returned useful data from deep space up to a fifth of the way to Venus and was never heard from again. The failure was made even more frustrating by the fact that the spacecraft may have performed its mission flawlessly, and that it was Soviet transmitters on Earth that had broken radio contact, never to be regained.

On 22 July 1962, the US tried for Venus, its first ever interplanetary mission. Five years after the start of the Space Race, a magnificent interplanetary spacecraft befitting of an all-conquering superpower was planned for the flagship mission. There was only one problem: it wasn't ready. As former JPL engineer John Casani relates: *"Mariner was to be a new design, really a second generation spacecraft that was specifically designed to go to Mars and Venus. But the design was not far enough along to be used for the Venus '62 opportunity. So we actually took a Ranger [Moon probe]."* In fact, the desperate engineers had just a week to plan something, anything, that would be ready before the deadline. They stripped the cameras from a spare probe originally designed to kamikaze into the Moon and refit it for a deep space encounter.

The hastily prepared Mariner 1 launched on an Atlas rocket with the hopes of the nation, and promptly veered off-course. A simple typo in the launch programming had added an unwanted minus sign, and America's first interplanetary spacecraft was deliberately exploded to smithereens in a self-destruct sequence without ever leaving Earth.

There was one last chance for the US to beat the Soviet Union during the Venus launch window and, just over a month later, the flight spare spacecraft, Mariner 2, made it into space. As the dizzy spacecraft plunged into interplanetary space, a NASA official outlined the agency's major concern that the spacecraft must operate unattended for nearly four months. Setting a precedent, a hastily modified Ranger spacecraft, whose original design hadn't even been able to reach the Moon, was expected to last many times longer than anything before it. Also, for the first time, mission controllers would have to learn the craft of controlling a spacecraft and solving problems on the fly. Problems were not long in coming.

Barely a fraction of the way to Venus, Mariner 2 needed to find a way to point its high gain antenna to its home planet with a critical Earth sensor. Ground controllers gave the command to 'find Earth', hoping to boost the fading signal from their spaceship as it flew ever further away. Mariner 2's radio transmissions kept getting weaker. Soon the signals would be too faint to track from Earth and Mariner 2 would be lost forever like its Soviet counterpart, Venera 1. Suddenly, one of the many Mariner miracles occurred. John Casani: *"And just a few days before it would have stopped being able to track, the signal strength jumped up to where we designed it and we were able to do the rest of the mission just fine with that Earth tracker."* Somehow, and not for the last time, Mariner 2 had mysteriously healed itself.

On 8 September, Mariner 2 started behaving crazily, turning on its gyroscopes and spinning out of control. The hapless spacecraft had been hit by the equivalent of a sniper's bullet in space, debris probably no larger than a grain of sand. Somehow, as controllers began to think they had once again lost their spacecraft, Mariner 2 performed another miracle and fixed itself. A month later, one of the two solar panels fizzled out. Project Manager Jack James rushed to find solar panels identical to Mariner's and performed tests in the Mojave Desert to try and pinpoint what went wrong, before his hapless spacecraft ran out of power. While in the desert, he had an idea. He deliberately shorted out the panel and found the symptoms to be almost identical to what was happening on Mariner. Just as orders to shut down the faulty solar panel were being sent to the spacecraft, the power problem healed itself. Just how these 'miraculous' healings occurred remains a mystery of the space age.

Finally, 110 days after launch, the power starved, overheating Mariner 2 somehow made it to Venus, as the US was recovering from the Cuban Missile Crisis. In its 35-minute encounter, humanity's last vestiges of a charming, tropical paradise for its sister planet were forever ripped away. Mariner found Venus to be literally as hot as hell, not just at the cloud tops but all the way to the surface. The slowly spinning planet could not even support a drop of water on its surface, where pressures were enough to crush a nuclear submarine. Sulfuric acid-laden clouds ceaselessly rained the poisonous liquid, which never reached the ground, where metals like lead and zinc could comfortably liquefy. Scientists were initially baffled as to why conditions on Venus could be so bad beyond human imagination, though a young scientist named Carl Sagan had finished a doctoral study that might hold the answer. His radical thesis was based on climate change. Venus, being closer to the Sun had suffered a runaway greenhouse effect as re-radiated heat was trapped by Venus' atmosphere. Eventually, any oceans an ancient Venus may have once had were boiled off forever, leaving a superheated, lifeless world. Sagan's theory was so popular that it has influenced generations of climate change experts, raising fears a similar fate might befall Earth if industry continues to pump carbon dioxide into our atmosphere.

Unable to stop, Mariner 2 screamed past Venus to enter a permanent orbit around the Sun. By the time the spacecraft that 'barely worked', according to a NASA official, finally stopped working on 3 January 1963, NASA had been given a sorely needed boost. At a time when agency executives were trying to justify their existence following a string of Ranger lunar probe failures, the US had operated a probe further from Earth than any nation before it. Looking back, former project manager Jack James, said: *"There will be other missions to Venus, but there will never be another first mission to Venus."*

Mariner 2 also had a profound effect on human imagination. Centuries of wistful dreaming was ripped away and pulp fiction visions of swamp monsters carrying away semi-clad females vanished, a process that would be repeated just three years later when Mariner 4 flew past a barren Mars. *"Venus Says No... The message from Venus may mark the beginning of the end of mankind's grand romantic dreams,"* remarked a New York Times article of the time. NASA Venus researcher and advocate, Dr David Grinspoon, wrote in Venus Revealed: *"Alas, no giant jellyfish, tree ferns, or dinosaurs."* The idea of the beautiful planet, long revered as a place of love and the undying world for mythic beings, was finally dead. Venus had ceased to be Earth's idyllic twin, and few artistic creations of the oven-like world have been made since Mariner 2's shock discovery.

The surprise success of Mariner at Venus gave NASA the confidence to try for a more difficult planet to reach: Mars. As a result, the follow-on probe meant for Venus in the 1964 launch window was instead re-tasked for the world's first successful flyby of the Red Planet. The US would not return to Venus until 1967 with Mariner 5, a re-re-tasked Mars flyby probe. Meanwhile, the Soviets had not given up on Earth's sister world. Instead of a flyby, Sergei Korolev, principal Soviet space architect, handing over to the Lavochkin Association, designed a probe, Venera 4, that would plunge straight into Venus' cloud tops. True to Korolev's style, the ambitious venture would pull the Soviet Union once again ahead of the US with the world's first ever landing on another planet. The problem was that no one really knew what lay beneath Venus' impenetrable clouds. Based on little else, Lavochkin engineers decided to be conservative and allow for a Venusian surface pressure ten times that of Earth. Also, considering the possibility of landing in a planet-wide carbonic ocean, Venera's antenna would separate from the main probe via a dissolvable sugar lock fixture. As Venera 4 dropped into the Venusian atmosphere on 18 October 1967, Venus proved the engineer's assessments were not conservative enough. At first, things went well as the first working spacecraft descended by parachute into the atmosphere of another planet. The British Jodrell Bank radio dish, responsible for scooping the Soviet's first Moon pictures the year before, tracked Venera's descent into Venus' atmosphere as temperature and pressure measurements went up, and up, and up. When transmissions ceased 94 minutes later, Soviet scientists were jubilant that their work borne of the socialist revolution had landed. Their celebrations were cut short, when further analysis of the data showed Venera 4 had imploded 25 kilometres above the ground.

Higher specifications were called for, and Anatoly Perminov ordered construction and testing of stronger spacecraft for Venus. Soon prototypes were being shoved into a giant pressure cooker test chamber, designed to simulate Venus. Initial tests didn't go well. Engineers placed a shiny spacecraft in the centre of the chamber and following testing, returned to find the chamber empty. They turned away, about to question colleagues as to who stole their spacecraft, when one of them noticed drops of shiny liquid on the chamber floor. The solidifying metal mass on the ground was all that remained of their hard work. Veneras 5 and 6 followed their predecessor into Venus during the next launch window, but again were crushed while still

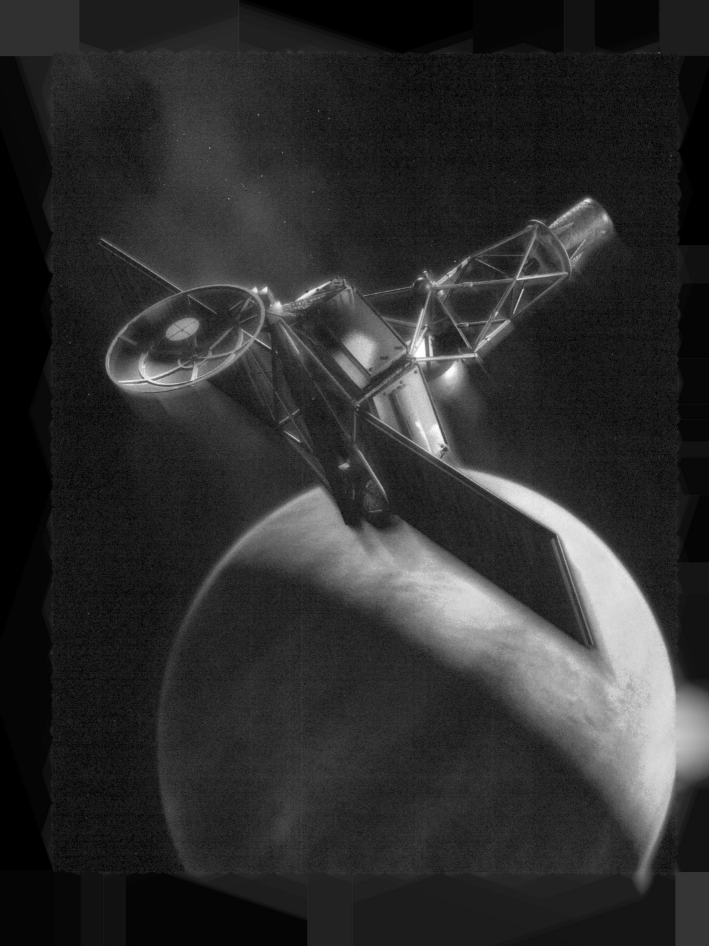

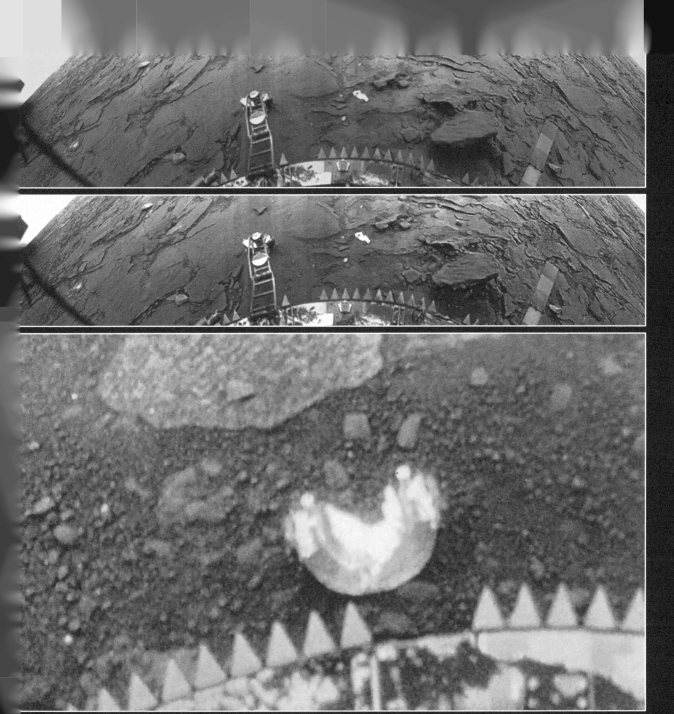

(Facing page) Mariner 2, the first ever artificial object to make a successful visit to another planet, passed by Venus in 1962. Mariner 2 returned hard data on a planet with temperatures hot enough to melt lead and surface pressures that would crush a military submarine. (Top) The Soviet Venera probes returned humanity's best views of the surface of Venus in the 1980's. Sunlight struggling through the sulfuric acid clouds casts a yellow hue (Ted Stryk). (Middle) Atmospherically corrected for Earth-like conditions, the Venera landing site was a flat lava plain. The penetrometer arm can be seen on the middle left. The arm unfortunately struck a lens cap, and was not able to obtain readings from the surface (Ted Stryk). (Bottom) A closeup view of a discarded lens cap resting on the surface of Venus. Although the Venera probes operated for minutes to hours, their corrosion resistant structure should last for decades to come. Firm plans to revisit the surface of Venus have

several kilometres above the surface.

Finally, an exasperated Perminov brought in some Soviet submarine experts, and decided to build a ridiculously strong titanium shell with barely enough room for even basic scientific instruments. On 15 December 1970, Venera 7 arrived at Venus. Its mother craft was left deliberately attached right up to the last second to maximise the amount of cooling it could deliver to its atmospheric cargo. Using parachutes trimmed to the bare minimum, Venera 7 dropped like a cannonball, stopping after just 35 minutes. Initially, Soviet mission controllers thought their latest Venus attempt had ended in a squashed pile of metal like the Veneras before it. Then, barely discernable above the radio chatter, a faint but measurable signal emerged, and pressure and temperature readings were stable. Venera 7 had made it! Lasting just over 20 minutes in the oppressive environment, Venera 7 announced to the world that the Soviets had succeeded in landing on another world.

Following their success with Venera 7 and later, Venera 8, the Soviets tried for another daring step; to take a picture from the Venusian surface. On 20 October 1975, Soviet scientists anxiously gathered around a dot matrix printer locked within a small room. The locked room ensured an overeager scientist couldn't 'accidentally' adjust the printer or grab the forthcoming printout prematurely. Venera 9 had landed minutes earlier, and the first ever picture from another planet slowly crawled out in black and white. What last hopes there might have been of a jungle world were dashed on the spot. The crude image showed a broken lava plain, dimmer than a cloudy midwinters day. The only liquid that had ever flowed here was lava from ancient volcanic eruptions.

Other Veneras followed, each managing to capture one or two images before their insides melted. In 1978, no less than 10 spaceships from two countries were sent to Venus. Along with the Soviet's Veneras, the US had chipped in with their fleet of Pioneer Venus explorers. Being revamped from budget cutbacks, the Pioneer Venus mission would send an orbiter and four atmospheric probes to Venus. As then NASA Ames Deputy Director Angelo 'Gus' Guastaferro wrote: *"It was clear that Venus would be investigated on a global scale for the first time, and that the entire planetary environment could be examined from in situ and remote sensing instruments."*

Venus was mapped, poked and probed as never before. The 1980's brought more Soviet spacecraft, this time to try and touch the Venusian surface. Veneras 13 and 14 came equipped with a penetrator designed to hammer the ground and provide much needed detail on what Venus was actually made of. Venera 13's instrument was dead on arrival and it was up to Venera 14 to try for Venus. Astronomical Society Australia member Igor Rozenberg described what happened next: *"The arm was not a normal robotic manipulator. It was just one sort; you deploy it and you can't move it any more. The issue was that they did not realize that this arm, which has a sensor on it hit the metallic cap of the camera! They managed to reach orbit, they managed to land, and after all this hell, then they had this mishap which they cannot actually touch the Venusian surface."* In one-in-a-thousand odds of failure, humanity had missed out on this chance of investigating the stuff of Venus.

Undaunted, the Soviets sent four more spacecraft to Venus. Veneras 14 and 15 used radar to map Venus so well that Soviet scientists boasted it would take 75 years before all the data could be processed. As it turned out, the development of the internet and the microcomputer revolution decreased the actual processing time by orders of magnitude. In 1986, Vegas 1 and 2 began a mission that was not matched for three decades. As the Vega spacecraft passed by Venus, they dropped off landers and balloons. The motherships continued to a historic encounter with Halley's Comet. The balloons, supplied by the French as a carry-on experiment, floated in the poisonous atmosphere for two days, surviving vicious air currents before their batteries ran out. For the Soviet Union, this was another first achievement (the first balloons sent to another planet) and final farewell. No other planetary spacecraft ever made by that country performed as successfully, and in just four more years the Soviet Union ceased to exist as well.

The results of mapping and sniffing the atmosphere began to paint a revealing and chilling picture. The spaceship armada reaffirmed that Venus was actually our sister world, being closer in size, mass and distance from the Sun than any other world. As scientists analysed the data in more detail, the resemblance became more disturbing. Early in the Solar System's history, Venus was even more like the Earth: as predicted by astronomers a century ago, Venus was covered in water. As both planets' surfaces began to cool from their formation; water, gifted from random acts of comets, caused primordial waves to lap the shores of both worlds. Even more tantalizingly, as with Earth, Venus might have given birth to life of its own. The abode of living creatures that Proctor envisioned in 1870 could have multiplied abundantly, but something began to go terribly wrong. The Earth, with its distance from the Sun and fast rotation, was able to recycle enough of its heat and carbon dioxide to keep climate change at bay. Venus wasn't so lucky. Its rotation happened at a walking pace and wasn't enough to drive the trade winds we enjoy on our own planet. Venus' oceans got hotter and began to evaporate. More of the Sun's heat got trapped beneath the ever-thickening clouds, instead of escaping to space. Once this feedback loop got going, there was no stopping it. Eventually, any water boiled away to be later ripped apart into atoms by the Sun and blown away by the Solar Wind. Whatever life there might have been on Venus had no chance; it was expunged completely, leaving behind an oven of a planet with the hottest surface temperatures of the Solar System. Could this be a warning for Earth, as industrialisation continues to spew carbon dioxide into our own atmosphere? In 1992, planetary scientist Donald Hunten summarised the Pioneer-Venus findings: *"There is no likelihood that the Earth will actually come to resemble Venus, but Venus serves both as a warning that major environmental effects can flow from seemingly small causes, and as a test bed, for our predictive models of the Earth."*

Although evidence of life on Venus was long gone, evidence for water might yet be found. The only way to find out was to send another mission. The Soviet Veneras mapped less than a third of the planet at high enough resolution. In order to see the detail that NASA wanted, the new mission would need to make better maps than available on Earth. On a high from the massive data haul of the Venera and Pioneer Venus missions, Stephen Saunders and his team of Venus explorers hurriedly began dreaming up a new proposal for a Venus mapping craft.

The only problem was budget cuts. The 1980's was a very lean decade for NASA, as the Space Shuttles swallowed most of the available funds. Undeterred, Saunders and his team looked for the equivalent of a fire sale to save their spacecraft, later named Magellan. Here was a dish left over from the Voyager missions, there was a spare computer from the Galileo Jupiter exploring spacecraft. Almost everything else was stripped off, including science. As NASA put it: *"The… spacecraft has only one science instrument: a radar sensor. This one instrument, however, will perform three important functions: collecting imaging data of the surface of Venus, acquiring altimetric data of the planet's topography, and measuring the natural thermal emissions from the Venusian surface."*

Innovation borne out of survival was also needed to make the one science instrument useful. A NASA press release explained: *"A large antenna on a spacecraft, however, would be expensive and difficult to manipulate. To solve this problem, the signals from Magellan's synthetic aperture radar (SAR) will be computer- processed on Earth so that they will imitate, or synthesize, the behaviour of a large antenna on the spacecraft."* Synthetic aperture radar would allow Magellan to resolve objects many times smaller than the best maps of Venus available at the time. That was assuming Magellan could actually get off the ground. Against all odds, the cobbled-together Magellan almost made it to the launch pad when again, disaster struck. In 1986, Challenger, the pride of NASA's Space Shuttle fleet, exploded mid-flight, killing all on board and grounding the fleet. It was not until 4 May 1989, after a safety review, that the Space Shuttle Atlantis launched and deployed the fleet's first ever planetary probe, Magellan.

Taking almost twice as long as most other missions before it, Magellan finally made it to Venus, conducted a critical orbit manoeuvre, and promptly broke down. After years of planning and waiting, engineers heard nothing but silence from their little spacecraft. NASA was just recovering from the Hubble Space Telescope fiasco, where a simple engineering defect lead to an almost fatal flaw in the billion-dollar telescope's primary mirror. The very real possibility of yet another NASA embarrassment was on the cards when suddenly, Magellan awoke from its seizure and started working. A month later than planned, and threatened with more budget cuts, Magellan began mapping Venus, swath by swath.

As Dr David Grinspoon wrote in Venus Revealed: *"Finally the fog was lifting from the surface of Venus. It was not a gradual clearing but a sequential peeling back of the clouds in thin strips from west to*

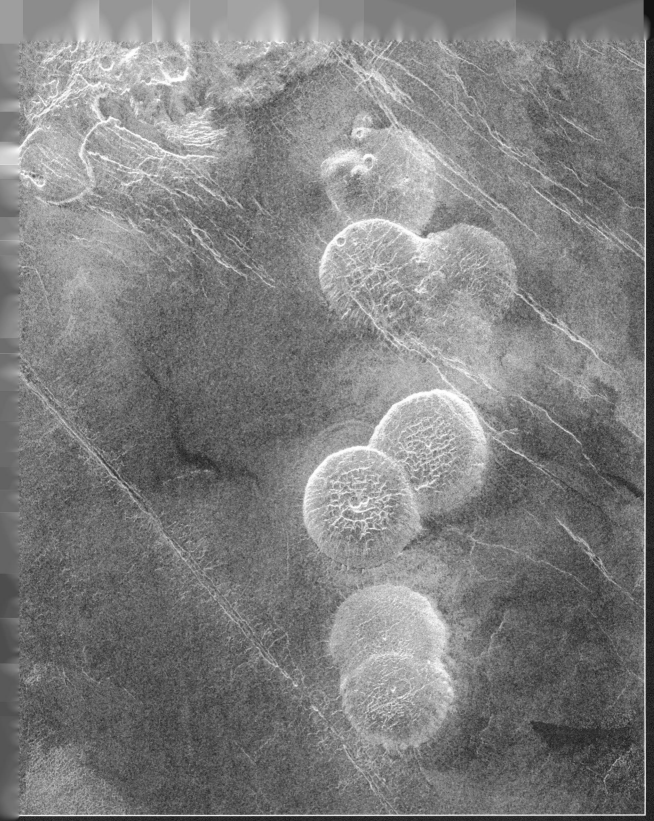

'Above) Pancake-shaped volcanic features were one of the bizarre landscapes discovered by Magellan. Scientists think Venus's dense atmosphere influences how volcanos work on the planet (NASA).

east. The Magellan images were so clear that we felt we were seeing Venus for the first time." The ultra-high resolution radar showed a world governed by the stuff of volcanoes. More than 80 percent of the surface is covered by lava flows and volcanic plains. Weird pancake features, and circular spider web-like formations joined fractures and faulting that spoke of a long history of eruptions. Along with lava, Magellan also uncovered dunes carved by Venus' ultra-dense atmosphere and snowfalls of metal. As with water snow on Earth, droplets of metal somehow condense out of the Venusian clouds and falls on the highland mountain peaks.

Magellan also looked for evidence of past water. Would fossilised shorelines or rivers be found by the spaceship's probing radar? Towards the end of the mission, some of the valleys, including Aphrodite Terra, a nearly 7,000-kilometre-long depression, were looking decidedly unlava-like. In a 1994 Icarus paper, author Jeffrey Kargel and his colleagues wrote: *"Venusian canali, outflow channels, and associated volcanic deposits resemble fluvial landforms more than they resemble volcanic features on Earth and Mars."* Could these strange valleys be the smoking gun of past water on Venus? Discussions raged almost as hotly as the planet itself, with the high resolution radar of Magellan, which had finally ceased operating in October 1994, not enough to solve the mystery.

Two decades on from Magellan, Venus has remained untouched by NASA. The Venus Exploration Analysis Group put it this way: *"'What happened to the water' is the question that has been the major driver of NASA's efforts to explore Mars in the last two decades... In contrast to this very healthy and scientifically productive set of missions to Mars, the Magellan radar surface mapper, launched in 1989, is the last dedicated NASA mission to our other planetary neighbour, Venus, where 'what happened to the atmosphere' is a paramount question also."*

The torch was passed to other countries. Europe's European Space Agency (ESA) took up the challenge of Venus, and on 9 November 2005 launched their "Mission of discovery" from Kazakhstan. Compared to the headline-grabbing Huygens probe landing on Titan, built by the same agency, the journey of Venus Express seemed mundane. The little spacecraft itself was a virtual copy of the Mars Express probe that had travelled to a more distant Red Planet two years earlier. This made design and operation cheaper and faster. The uneventful entry of Venus Express into orbit gave ESA scientists more than they bargained for. A massive extra fuel margin meant the planned three-year mission was extended no less than five times, eventually losing contact in 2014. Flying low enough to scrape the Venusian atmosphere, ESA discovered super vortices: massive, cyclone-like weather formations focused on the planet's poles. As ESA wrote on their website: *"Several planets in the Solar System, including Earth, have been found to possess hurricane-like vortices... but none of them are as variable or unstable as the southern polar vortex on Venus."*

Along with discovering high altitude snow that never makes it to the ground, and an ozone layer, Venus Express tried to answer a question that could change the way humanity looked at the cosmos. If Venus once held life, where did it go? Dark streaks, only visible in ultraviolet, show in images of the ultra-thick atmosphere and have intrigued scientists. Once thought to be the actual Venusian surface by pre-space age astronomers, another, more exciting possibility has emerged. What if, instead of boiling alive along with the rest of Venus on its trip to hell, primitive life instead ascended to the heavens? Could the first evidence of life outside Earth not be found in the rust red sands of Mars, but instead be discovered floating in the Venusian clouds?

University of Wisconsin Senior Scientist Sanjay Limaye imagined the last Venusian survivors existing in a narrow sweet spot, absorbing ultraviolet from sunlight and impervious to sulfuric acid: *"There are questions that haven't been fully explored yet and I'm shouting as loud as I can saying that we need to explore them... I cannot say there is microbial life in Venus clouds but that doesn't mean it's not there either. The only way to learn is to go there and sample the atmosphere."*

Unfortunately, as Venus Express dived into the Venusian atmosphere for the last time in December 2014, Sanjay's questions about life on Venus remain unanswered, despite Japan launching its first ever mission to Earth's wayward twin in 2010. Unlike the 'routine' Venus Express, Japan's Akatsuki missed the planet, becoming a small planet of its own for five years before finally reaching its goal. *"It's the dawn of a new day for science at Venus,"* said NASA planetary scientist director Jim Green. Although Japan Aerospace Exploration

Agency (JAXA) had overcome adversity at Venus, an atmospheric sampling mission will probably be needed before we can finally know the true nature of Venus' clouds. Is Venus really a twin of Earth, hosting life in its clouds, or is the most Earth-like planet in our Solar System a sterile, inhospitable hunk of rock?

For tiny Mercury, the Space Age did not start until 1973. By this time, Proctor's dream of a moisture-laden world had long since boiled away. Increasingly accurate estimates of its surface temperature so close to that of the Sun made the existence of life-giving liquid water, or even an atmosphere, unlikely to 20th Century astronomers. Additionally, Mercury was (and remains) a difficult destination to travel to. Extra powerful and hence prohibitively expensive rockets would be needed to slow a spacecraft down and fall inward to the Sun. More thrust would need to propel the space ship far enough to reach this tiny world. The nearer and larger targets of Venus and Mars were more than enough for the early Space Race to concentrate their efforts on.

After decades of speculation, NASA finally revealed the key to unlocking a cheap journey to Mercury. A theoretical concept, called the gravity assist principle, meant that a prospective explorer could use the gravitational fields of other planets to change their speed and trajectory. Using the host planet's energy in this way would mean using cheaper rockets, making the mission more viable. The only two drawbacks of the principle were that no one had actually tried it before, and tiny navigational errors at the host planet could mean missing the desired target by millions of kilometres. However, NASA was confident that existing guidance systems would be adequate for the task and planning for a gravity assist Mercury mission began.

Computer-tested flight trajectories were compared at the JPL and favourable launch windows were identified in 1970 and 1973. The 1973 launch window was chosen as it would be possible for Mercury to be revisited following the first fly-by. The proposed spacecraft would first travel to Venus, and be slowed down by its gravity to fall sunward until it reached Mercury. Fuel permitting, it would then use its twice-Mercury-orbital-period to revisit the planet for a second, and maybe a third time. Even then, the proposal was not approved until 1969 due to budget cuts and JPL director William Pickering was given just USD $98 million to cover the whole project. In comparison, the Viking missions to Mars built over roughly the same period cost over USD $1 billion. Moreover, the team had just four years to get the mission off the ground.

To save costs, Pickering chose to use the tried and true Mariner spacecraft design, already proven in previous Mars and Venus encounters. Spare parts left over from other missions were used, a process repeated decades later for the Magellan mission to Venus. In a break from tradition and learning from previous missions, planetary scientists sat down with engineers to help plan their spacecraft from the start.

Mariner 10 launched within a few thousandths of a second of its launch window opening on 2 November 1973, and began flying backwards to fall inwards to the Sun. Despite initial heater problems, the spacecraft captured test images of the Moon. According to NASA's The Voyage of Mariner 10: *"Experimenters were enthusiastic about the excellent quality. The Moon pictures recorded objects a mere 3 km... across."*

Since the pictures to be returned from Mercury were expected to be of three times higher resolution than those of the Moon, there was good reason for excitement. At last, it seemed, mankind would have a chance to resolve those dusky markings on the innermost planet, those indistinct features that earlier astronomers had interpreted as Mars-like, even erroneously with linear 'canal' type features."

Following a harrowing mid-course correction, whose omission would have lead Mariner 10 to be lost in space, the only scheduled visitor to Mercury lost track of its guide star and began to have power trouble, threatening the mission. In a press release NASA stated: *"... the most significant and ominous power-related problem did not occur until January 8, 1974, when the spacecraft automatically switched from its main to its standby power chain. This automatic switchover was irreversible: it was of concern primarily because of the possibility of a fault common to both power circuits causing the backup power circuit to fail also and thus raising the possibility of the mission's being ended right there."* The spacecraft had also wasted nearly a fifth of its fuel during this time, and it had not even reached its first destination.

As with its predecessor, Mariner 2, a combination of careful nursing from mission control and mysterious 'healings', Mariner 10 managed to make it to its rendezvous with Venus on 5 February 1974. Although not the primary target of the mission, Mariner 10 captured some of the best images of Earth's twin planet, and also investigated the ultraviolet absorption layer in the atmosphere.

(Above) Lightning strikes illuminate the otherwise impenetrable clouds of Venus. Even on a calm day the Venusian sky rumbles with lightning. Below, an orange cast colours the volcanic landscape during the 24...

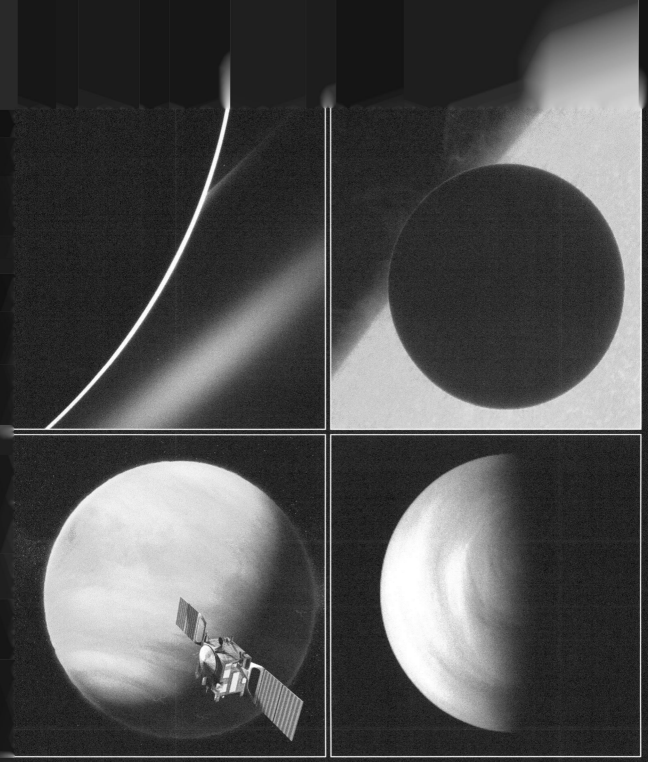

(Top left) Venus is bright enough to be seen from Saturn. Looking back towards the inner solar system, Cassini imaged Venus as a small dot in the haze of one of Saturn's rings. (Top right) A rare transit of Venus was imaged by a solar observing satellite. Venus's atmosphere forms a ring around the dark side of the planet from refracted sunlight (NASA). (Bottom left) ESA's Venus Express analyzed the planet for nearly 10 years before plunging into the atmosphere it had been sent to investigate. (Bottom right) A view of the Venusian pole from

Losing some of its velocity to Venus, Mariner 10 fell further towards the Sun, on track for Mercury. As Venus and its active atmosphere receded in Mariner 10's cameras, the spacecraft again lost a lock on its guide star, wasting more precious fuel. Tiny meteoroids were flashing past the star sensors, which confused them for navigation stars. As Mariner 10 closed in on Mercury, these problems increased tenfold. Scientists and engineers worked together to carry out repair-on-the-go procedures to save their spaceship. A version of this process would be repeated years later for the Voyager missions to the outer solar system. Somehow, despite the odds, Mercury's first space-age visitor reached the planet on 29 March 1974.

First spots, then crater rays, then finally craters were revealed as Mariner 10 first closed in, then flew past the inner most planet. Scientists barely kept up with the stream on incoming images, hurriedly assembling them into photomosaics. The real Mercury began to emerge, piece by piece. One particular photomosaic caused a collective gasp to rise up from the image processing room. Half-hidden in shadow, Caloris Planitia, a massive circular basin a quarter the size of the whole planet dominated the image composite. Mariner was showing the Solar System to be a violent neighborhood. Until Caloris, many had thought Earth's heavily-cratered Moon to be an oddity, with its proximity to Earth somehow making it a shooting gallery for meteors. Mercury was showing this was not the case, and the whole Solar System had been heavily bombarded in its early history. Whatever had smashed into Mercury to create Caloris was so big that the shock waves had raced right around the planet to create hummocky terrain on the opposite side.

As Mercury receded behind Mariner 10 from the first of its three encounters, JPL Director Bruce Murray reflected on its findings: *"Mariner 10 returned almost 3,000 pictures of the surface of Mercury. These features and these pictures are like the pages of a history book. Now we can compare our similarities and differences and try to recognize our family relationships among the terrestrial planets. Are we cousins or brothers or are all of us bizarre strangers that happen to inhabit the same portion of the Solar system?"*

Mariner 10 found cratered Mercury to have a magnetic field, and that the tiny planet had gotten even tinier in its history. Here and there were craters cloven in two by enormous cliffs that scraped the sky. Somehow, Mercury had contracted and its hard outer shell buckled, cracking at various weak points. Yet, despite all the mission had accomplished, mysteries remained. Due to the quirks of celestial mechanics, Mariner was only able to image half of the surface of Mercury. Potential Mercurian cities could have resided on the dark sides of the planet, for all NASA knew. Additionally, why did Mercury, little bigger than Earth's Moon, have a magnetic field? According to reasonable thinking, a planet that size should have become stone cold and lost its magnetic field long ago, just like the Moon, or Mars.

Mariner 10 would not be able to provide any more answers. Out of fuel, the pioneering spacecraft was commanded to turn its transmitter off on 24 March 1975. Still circling the Sun in its solar orbit, the now derelict Mariner continued to revisit Mercury in the decades that followed, as a silent ghost ship. Writing in the Icarus Journal in 1975, astrophysicist David Morrison remarked: *"... it seems certain that the Mariner data will continue to be analyzed for many years to come, and that this planet is now firmly fixed in both public and scientific consciousness."*

The years that followed certainly saw Mariner 10's data shared around the planetary science community as the pre-Space Age telescopic hazy dot had finally become a real planet to humanity. Photographs of Caloris basin and other parts of the planet were published in astronomy books. Then forty-year-old Robert Strom was one of those actively analyzing Mercury data. He had worked hard during the Mariner 10 mission, supporting the cantankerous spacecraft through its misbehaviors. Like the rest of the now disbanded team, he looked forward to a follow-up mission while Mercury was firmly fixed in both public and scientific consciousness.

The 1970's were a busy time for space exploration as NASA prepared for its first ever landing on Mars in 1976, and sent the Voyager spacecraft to the outer Solar System. Mariner 10 and Mercury were soon forgotten by the public. The problem, according to Robert Strom, was that the pictures from Mercury didn't seem interesting enough: *"These images did not reveal enormous shield volcanoes, deep canyons and huge flood channels as occur on Mars, nor did they show the huge fracture belts, large volcanos and peculiar surface features that occur on Venus. Most people thought it was a dead Moon-like planet, and therefore, not as exciting as the other bodies in the solar system and a low priority for further exploration."*

Mariner 10 also proved getting to Mercury was expensive. The venture had cost a billion US dollars in present-day money, and the financially lean decades that followed were no times for adventures to a Moon-like planet. A whole generation was to pass before the Solar System's innermost planet would again receive a visitor from Earth.

Sean Solomon, who would later become principal investigator of NASA's next mission to Mercury, was one who took up the challenge of planning a return mission. Instead of flying past Mercury, as Mariner 10 had done, the next mission would orbit. Solomon knew this was no small feat, as the amount of fuel needed by a spacecraft to slow down enough to be captured by small Mercury would be too much to carry. In fact, even if the spaceship was all fuel and nothing else, it would still not be enough for the task.

In 1992, a breakthrough discovery, inspired by the planet-hopping saga of the Voyagers, came from a JPL engineer. The engineer realised that by flying past Venus and Earth a number of times, a Mercury spaceship would lose enough speed to orbit Mercury. The trade-off would be that the mission would need to be in space a lot longer, years in fact, before finally reaching its goal. The other cost was financial, a billion dollars, by 1990's estimates, owing to a planned launch from the Space Shuttle. The idea was dead on arrival.

Four years later, the Space Shuttle program faced the devastating crisis with the loss of Challenger that had affected the Magellan Venus project. Although tragic, some follow-on effects of the Challenger disaster actually benefited space exploration. Up to 1986, the Shuttle was being forced as the primary method of launch for NASA, for political as well as scientific reasons. Commercial payloads, as well as NASA satellite and spacecraft payloads, were needed to pay for the Shuttle. After all, why have a Space Shuttle at all if people weren't going to use it to launch payloads into space? Following Challenger, the Shuttle fleet was grounded, along with high profile and expensive space missions, such as Galileo to Jupiter. Facing budget overruns themselves, some service providers chose instead to use cheaper rockets to launch their missions into space. By this time the Mercury mission, now called Messenger (Mercury Surface, Space Environment, Geochemistry, and Ranging), was again proposed as a mission. Launch from the Space Shuttle, which by now focused on building the International Space Station, was no longer on the cards. Messenger's price tag dropped to a quarter of its original estimate, and in 1999 the mission was approved.

As Messenger began to take shape, other issues besides costs arose. Project Manager Ralph McNutt and principal investigator Sean Solomon realised the significance of sending the first spacecraft in 30 years to visit Mercury. The decades since Mariner 10 had raised lots of questions about Mercury and it seemed almost everyone had a pet scientific instrument they wanted to include on the mission. Mercury team member Sean Solomon stated: *"The biggest challenges we faced on the mission was that we had a quite a [diversity] of scientific objectives, we had a payload of seven instruments; they had different requirements for pointing... different types of data. We had some severe constraints on the attitude of the spacecraft because we had a sunshade that had to be between the spacecraft and the sun at all times. So the biggest challenge was to figure out how best to take the observations with all of the instruments that would maximize the scientific return across all of the objectives... It took technology more than 30 years, from Mariner 10 to Messenger, to bring us to the brink of discovering what Mercury is all about."*

On 3 August 2004, Messenger left Earth for its roundabout route to Mercury. Passing Earth again in 2005, and Venus twice in 2006 and 2007, engineers used the extra time forced on them to do shakedown tests of their craft. 614 pictures were snapped of Venus in a dress rehearsal for the real thing to happen in 2008. As the spacecraft sped towards its rendezvous with Mercury, JPL Space Department Head Michael Griffin stated: *"With Messenger on its way to Mercury, the reality is sinking in that in a few years, we will see things that no human has ever seen..."*

Messenger flew past Mercury three times between 2008 and 2009 before finally settling into orbit in 2011. For one Mercury enthusiast, by then in his 70s, the event was literally a dream come true. Robert Strom: *"When Messenger went into orbit about Mercury on March 18, 2011, I was overjoyed. I had waited almost 40 years for that moment."*

Messenger transformed Mercury from a dull rock with craters into a far more interesting place. As Michael Griffin had predicted, humanity got its first look at the whole of Mercury. While no ancient

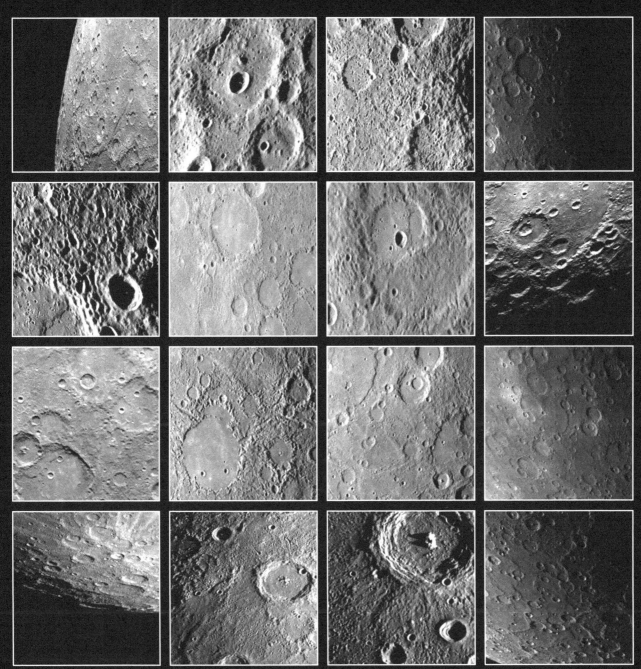

(Facing page) By the time the space-age caught up with Mercury, most scientists understood the planet to be an airless world scorched by the sun. Theoretically, astronauts visiting Mercury would only need reflective spacesuit material and insulated footwear to explore the surface (David A. Hardy). (Above) Mariner 10 revealed Mercury to be a cratered world, lacking the expansive lava plains found on our Moon (NASA).

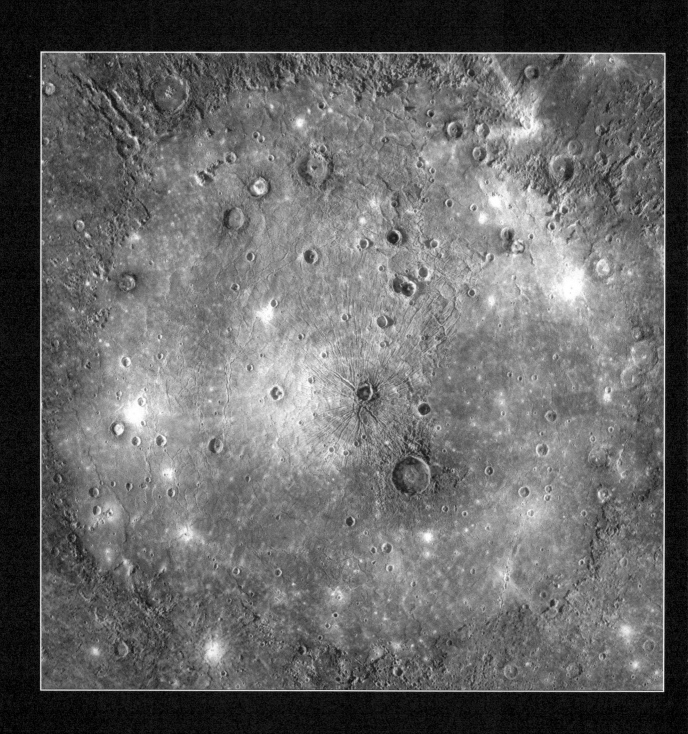

(Above) NASA's Messenger probe provided a long overdue second look at Mercury. A strange spider-like crater lies near the centre of this false colour image (NASA).

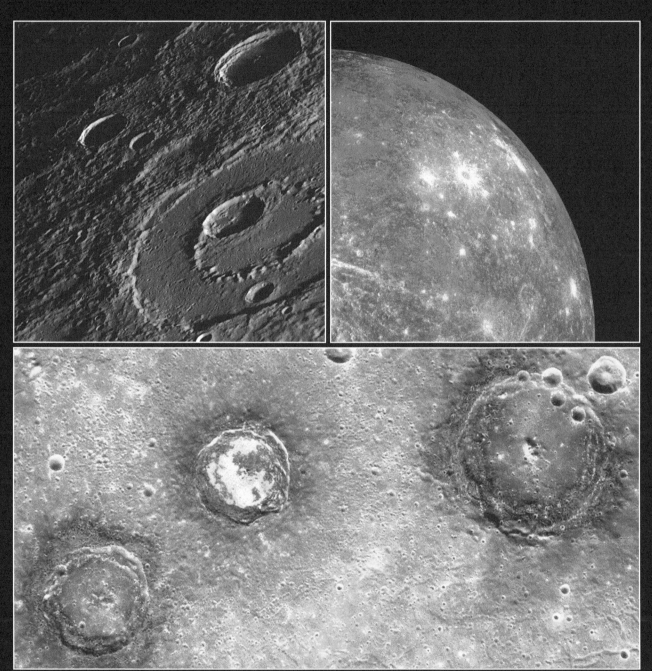

(Top left) A triple crater emerges from Mercury's shadows, imaged by Messenger. (Top right) The Caloris Basin, shown here in false colour, is the largest impact basin on Mercury. The impact that created it was so great that shockwaves created chaotic terrain on the opposite side of the planet. (Bottom) False colour imagery of Mercury's craters, showing different materials making up the crater walls (All images NASA).

Contemporary artwork of the surface of Mercury, based on the latest Messenger results. The enlarged sun illuminates ridges created by Mercury's shrinking crust (Mark A. Garlick).

civilizations were seen, Messenger uncovered other exciting discoveries. For starters, Messenger investigated the mysterious shrinking of the planet. Using cameras far more sensitive than the primitive vidicon instruments on board Mariner 10, Messenger discovered that Mercury was up to 15 kilometres smaller in diameter than when it had first formed. The incredible shrinking process seemed to be still ongoing, as some of the crustal cracks on Mercury were only a few million years old. Smithsonian senior scientist Tom Watters said of the activity: *"There are similar examples of this on Earth involving both oceanic and continental plates, but this may be the first evidence of this geological process on Mercury."* Why the small Mercury was still geologically active has yet to be solved.

Sean Solomon and the rest of the science team also found more evidence of an active Mercury. Massaging the Messenger images to their limit, the team saw what looked like volcanic vents, dotted throughout the landscape. Perhaps Mercury was a world of fire after all. Sean Solomon: *"We're building up a catalog of probable volcanic centers... that's a surprise."*

Messenger also made a discovery that initially seemed to be the last thing anyone would have expected on a planet this close to the Sun: ice. Flying over the poles of Mercury, Messenger peered deep into the darkness. Unlike Earth, Mercury does not tilt on its axis in its journey around the Sun, and the bottoms of some of the polar craters have never seen sunlight. Covered in a protective organic matter, reserves of frozen water, probably deposited at the dawn of the solar system, have lain preserved from being scorched for eons. Sean Solomon: *"Those polar regions, I think, are calling out to people like [NASA Planetary Science Division Director] Jim Green and saying, 'Send us another spacecraft, we have more stories to tell."*
Discovery by discovery, Mercury was regaining the mystique bestowed on it by pre-Space-Age astronomers.

After an amazing adventure, and years of fighting the gravity of the Sun and Mercury, Messenger ran low on fuel. Yet the little spacecraft had one final, abrupt task to complete. Science team member Thomas Zurbuchen explained: *"We're at the end of a really successful mission, and we can't do anything anymore to stop it from doing what it naturally wants to do. The sun is pulling on it. The planet is pulling on it. It's just physics. It has to crash."*

NASA followed the tradition of giving spacecraft a voice through Twitter: *"Well, I guess it's time to say goodbye to all my friends, family, support team. I will be making my final impact very soon."* Using the last of its fuel, Messenger fired its rocket for the last time and prepared to crash into the planet it had studied for so long. On 30 April 2015, transmitting data until the last moment, Messenger's mission came to a dramatic end as it slammed into Mercury with the speed of a meteor and gouged a brand-new crater of its own.

Although the spacecraft itself was forever silenced, scientists would spend years sifting through the data to try and understand Mercury. NASA Science Mission Directorate Associate Administrator John Grunsfeld: *"The MESSENGER mission will continue to provide scientists with a bonanza of new results as we begin the next phase of this mission — analysing the exciting data already in the archives, and unravelling the mysteries of Mercury."*

In its own way, Messenger had rewritten the textbooks about Mercury and pulled the little planet from obscurity to a world of interest. Scientists are still processing data about the Solar System's innermost planet, trying to understand the nature of its magnetic field, and even why the small world has an atmosphere, as discovered by Messenger. To be sure, the wheeze of Mercurian gas didn't hold a candle to the life-giving mantle of Earth, but it was an atmosphere all the same. Also, why did Mercury, like a snowball in hell, have ice in the poles? As is the case in science, Messenger, like Mariner 10 before it, has both answered old questions about Mercury and raised new ones. In 2024, a joint European and Japanese mission, BepiColombo, will travel close to the fires of the Sun and further investigate the Solar System's innermost planet.

The same interest has not followed Venus. Somewhere in the ochre, overheated plains of our sister world resides a unique artefact. Derelict but structurally intact, the yellowing bulk of Venera 14, the last probe to successfully reach the forbidding surface, rests to this day. The now lifeless camera eyes stare out across the shimmering lava field and look to an uncertain future. Decades after Venera 14, there are no firm plans to darken the Venusian skies with another lander. A dwindling group of Venusians, of which David Grinspoon is an advocate, has certainly been busy creating proposals for future landers. Better technology would also make a lander that could live for days or weeks, rather than broiling alive in just minutes, very possible.

In 2017, the light of Venus shone bright for the intrepid group, as not one but two Venus missions made it through the first round of proposals. Former NASA Chief Scientist Ellen Stofan saw Venus as an important goal in the study of planets in other solar systems: *"We're in [a] phase of looking at these extrasolar planets and saying: Now wait a minute—Venus has something really important to tell us about why the Earth is habitable and why is Venus not."*

Sadly, even Stofan's voice was not enough. To her surprise, and the bitter disappointment of the Venus group, neither proposal made the final cut. Starved of new mission data, the aging Venus research group would find it ever harder to attract new research. This has come at a time where understanding of planetary climate change is a critical priority for most governments. Stofan: *"My huge concern is that the expertise is still there but it's going away. If we have no Venus missions it will definitely be gone. We will have lost something. We will have lost this capability and it's really unfortunate."*

Whether the US or another country such as Japan will try for the surface of Venus is uncertain. For now, the forlorn hulk of Venera 14 remains as a remainder of a time when one of the most hostile environments in the solar system was explored.

(Above) The last earthly visitor to the surface of Venus, Venera 14, stands a forlorn watch over the desolate landscape. Even though the spacecraft stopped working long ago, there is little on the outside structure of the lander to crack or melt in the Venusian heat.

(Above) Early artwork illustrating a steamy Venusian jungle (David A. Hardy).

Of all the planets in the Solar System, only Mars rivals Venus in spawning a host of sci-fi imaginations of exotic aliens. Also like Mars, the space age showed how a world hyped up by science-fiction turned out to more or less be the complete opposite. The presence of ultra-bright clouds of Venus' atmosphere made it a prominent and beautiful world to the ancients. The third brightest thing in the sky so captivated the Roman Empire that it was named after their goddess of love and fertility. Meanwhile, the ever-present clouds which perpetually hid the Venusian surface from the gaze of Earth-bound telescopes, led to fertile imaginations populating the planet with steamy jungles and strange creatures. To 19th Century astronomers, such as Proctor, the similarities of Venus to Earth were clear: *"... in size, in situation, and in density, in the length of her seasons and of her rotation, in the figure of her orbit and in the amount of light and heat she receives from the sun, Venus bears a more striking resemblance to the earth than any orb within the solar system."*

      A theory in the early 20th Century suggested that the Sun was gradually cooling down, giving each planet in our Solar System the ability to harbour life. A hotter early Sun once supported a lively Mars, life on Earth was maturing while Venus was just starting up. A warm, wet world, fuelled by erroneous readings of atmospheric water by early spectrometers, was the suggestion of Nobel Prize winner Svante Arrhenius: *"[E]verything on Venus is dripping wet... A very great part of the surface ... is no doubt covered with swamps corresponding to those on the Earth in which the coal deposits were formed... Only low forms of life are therefore represented, mostly no doubt, belonging to the vegetable kingdom; and the organisms are nearly of the same kind all over the planet."*

(Above) An alternative pre space-age view of a drier, desert-like Venus (David A. Hardy).

As late as 1955, Russian scientist G.A. Tikhoff wrote: *"Now already we can say a few things about the vegetation of Venus. Owing to the high temperature on this planet, the plants must reflect all the heat rays, of which those visible to the eye are the rays from red to green inclusive. This gives the plants a yellow hue... I in my turn wish to add that here a certain part may be played by the vegetation of Venus. Thus we get the following gamut of colours: ...on Venus where the climate is hot the plants have orange colours."*

Other authors, such as Lucien Rudaux and Gerard de Vaucouleurs, were more cautious, knowing that there was no hard evidence to support any theory of what lay beneath the Venusian clouds: *"The truth of the matter is that we still simply do not know, and whether it will be possible to find out within the next few years remains to be seen."*

Effectively given carte blanche from the astronomy of the time, sci-fi stories such as 'The Great Romance' (anonymous, 1881), 'Journey to Venus' (Pope, 1895) and 'Space Cadet' (Heinlein, 1948) were set in a hot and swampy Venus. In these stories, explorers visited Earth's twin world to encounter prehistoric-style aliens or utopian Venusian societies. Olaf Stapeldon's 'Last and First Men '(1930) took a more horrific twist. In this story, people from a dying Earth committed genocide on Venus' intelligent, aquatic life and replaced them as the dominant species on the planet. Other great story-tellers, such as Ray Bradbury of Martian Chronicles fame and Isaac Asimov, set their writings on a warm, wet Venus.

What of the space artist? The paucity of scientific measurements of Venus' surface conditions gave those attempting to visualise a realistic impression of the surface of Venus almost as much freedom as the sci-

(Above) Contemporary artwork of Venus based on information from the space age (William K. Hartmann).

fi writers. To those living through the transition of the Space Age perceptions, artwork of Venus evolved over time. Bill Hartmann summarised his experience of Venus: *"My recollections from the early 50s are that before that time it was widely assumed that the massive Venus clouds were (like Earth's clouds) formed from water vapor, so that the Venus was a world of extreme rainfall."*

"Another model, more correct, illustrated by Bonestell in 'Conquest of Space' as early as late 1940's, was that water on hot Venus would evaporate, leading to a hot, dry desert under the clouds." Bill Hartmann also remembered spectroscopic measurement s around the 1950's that discovered most of Venus' atmosphere was carbon dioxide. *"...The 'rainy-world' model died out. The next big discovery was the thermal infrared data that Venus temperatures were not around 140 degrees Fahrenheit — which is what you get by moving Earth to Venus's orbit — but were rather in the 700-900 range. Carl Sagan, in his early work, was involved in the recognition that this was caused by the greenhouse effect."*

The artwork on the preceding pages reflects the evolution of humanity's perception of Venus. In 'Venus Primeval Earth', a young David A. Hardy painted the wet, steamy Venus that was popular with many sci-fi writers before the Space Age. As more reliable surface measurements revealed a drier Venus, Hardy painted a dry dustbowl of a planet. His later painting illustrated a wind-swept desert world, where rocks have been sandblasted into fantastic shapes over the eons. Apart from the enlarged Sun, Hardy's painting could almost show a location on present day Mars, where wind is currently the most active force. Millions of years of wind

(Above) Billions of years ago, Venus, like Earth, may have hosted seas of liquid water before boiling off.

action have moved sand dunes across the planet. In fact, the North Pole of Mars is totally surrounded by a massive sea of dunes.

A painting by Bill Hartmann (preceding page) illustrates humanity's current understanding of the surface of Venus. A bloated Sun struggles in a futile attempt to break through the perpetual sulphuric acid-laden clouds. The ground, consisting of shattered volcanic rocks, roils in an atmosphere superheated to over 460 degrees Celsius. The mostly carbon dioxide sky weighs down on the surface with enough pressure to crush a military submarine. The distant Maxwell Montes, the highest mountain of Venus, shimmers in the distance. The radar mapper on board the US spacecraft Magellan discovered that snows of metal fall on Venus. The alien snow, probably lead sulphide, falls on the peaks of Venus' tall mountains, such as Maxwell Montes.

Perhaps not surprisingly, sci-fi stories about Venus dried up, along with any idea of a twin planet of Earth filled with life. Sadly, artwork of the planet has also declined, as the attention of the Space Age shifted to Mars and other parts of the Solar System. Recent theories suggest that, along with an early Mars, primordial oceans, possibly containing life, once lapped Venusian shores. The presence of water on early Venus remains an open question. Perhaps one day, using current technology that would allow landers to last for days, or even weeks on the Venusian surface, hard evidence of water on early Venus might be found. If evidence of past or present life, escaping to the cooler reaches of Venus' upper atmosphere, were found, the planet might regain something of its romantic past.

WORLDS OF GAS AND ICE

Knowledge of the wanderers, as the ancients called them, is almost as old as humanity itself. One of the brightest of these wandering stars was called Jupiter, after the Roman king of the gods. It was also Jupiter that caused a revolution in western civilization's understanding of our neighbouring worlds. From the ancient's viewpoint, all stars, planets and the Sun seemed to rotate about the central Earth. This view, theorised by Aristotle and adopted by the Roman Catholic Church, drove astronomical understanding for centuries. Crystalline spheres, set in an ethereal element, serenely circled the central Earth in this idealistic view.

By the 15th Century, however, observations of eccentricities of the Solar System's motion, particularly by Copernicus, was challenging this simplistic perception of the universe. These were dangerous times, as the powerful Catholic Church opposed any challenges to its authority and teachings. Copernicus refused to have his radical astronomy work published until after his death in 1543. Scholar Giordano Bruno was burned at the stake, in part for suggesting life on other stars in 1600. Just a decade after Bruno paid the ultimate sacrifice, an Italian Catholic astronomer and scientist made a discovery that would revolutionise society and push astronomy to a secular science. Galileo Galilei pointed his telescope to the king of planets, and discovered Jupiter was not alone. Four firefly-like points of light danced around the massive Jupiter, sometimes disappearing and reappearing some time later. With astonishment, he realised that these little stars were moons, orbiting Jupiter as our Moon orbits the Earth.

This discovery, along with observation of phases of Venus, only made sense if the Sun was at the centre of the Solar System, and all other planets, including Earth, revolved around it. Galileo was later tried and forced to recant his theories of a Sun-centred, or heliocentric, Solar System, but his work inspired more observation-based work. Over the next century, discoveries by Christiaan Huygens, Giovanni Cassini and others, furnished such a mountain of evidence for the heliocentric Solar System, that the Earth-centred theory was no longer tenable to serious thinkers.

During these times, almost nothing was understood about the environmental conditions of the outer Solar System. The planets were thought to be more or less bigger versions of Earth. Even in the 1900's, many thought Jupiter to have a solid surface. In 1850, American physicist and astronomer Professor Denison Olmsted wrote: *"The belts are not ranges of clouds, but portions of the planet itself, brought into view by removal of clouds and mists..."*. Artwork by Chesley Bonestell depicted icy mountains spewing ammonia lava under dense clouds.

This view began to change. Using ever-improving telescope equipment, astronomers began measuring the times it took for different parts of Jupiter's surface to rotate around the planet. They did not line up, as a fixed surface should have. Irish astronomer R.S. Ball was one of those to understand the true nature of Jupiter: *"The conclusion is irresistibly forced upon us, that when we view the surface of Jupiter we are not looking at any solid body."*

Jupiter's 'surface' was discovered to be an unending series of storms and tempests. One of these, discovered as early as 1665, is the Great Red Spot. Named after its colour, this persistent feature was seen on-and-off in the southern latitudes of Jupiter, and continually observed since 1879. Early astronomers theorised the Great Red Spot, so large that two Earths could comfortably be placed side by side within it, to be an enduring cloud or, according to astronomer Martin Davidson: *"The Great Red Spot may, therefore, be a solid body floating in an ocean of permanent gases."*

During the time of the first telescopic discoveries in the 1600's, one of the dimmer wanderers, named after the lesser Roman god Saturn, came to the attention of Galileo. He observed what he thought were 'ears' that appeared and disappeared near the orange blob. In 1655, Christian Huygens, using a better telescope than Galileo, was able to refine the 'ears' to a ring surrounding Saturn. William Herschel, the astronomer who

(Facing page) The Cassini spacecraft is caught in a ray of sunlight as it flies over Titan, one of Saturn's moons. Cassini spent 13 years exploring the Saturnian system, returning breathtaking images.

would discover the planet Uranus, wrote: "It is surrounded by a thin, flat, ring, nowhere touching, inclined to the ecliptic." In 1675, Giovanni Cassini discovered a division, later named after him, in the rings. In 1860, postulating what the rings would look like to an observer located in Saturn's clouds, Professor Olmsted wrote: *"The rings of Saturn must present a magnificent spectacle from those regions of the planet which lie on their enlightened sides, appearing as vast arches spanning the sky from horizon to horizon."*

Theories abounded as to what the rings were actually made of: whether they were some kind of solid disc, or made up of tightly bound rolling pin like rubble, endlessly turning around Saturn, or even a swarm of tiny moonlets. Discoveries in the 1700's of the Crepe ring and additional gaps raised doubts about a solid mass. R.S. Ball wrote: *"... every part of that structure would be pulled forcibly towards [its] surface, and thus the materials of the arch, if it were a solid body, would be compressed with terrific force."*

Finally, in 1857 James Clerk Maxwell, author of Maxwell's equations describing electromagnetic radiation, used his mathematics to show: *"The only system of rings which can exist is one composed of an indefinite number of particles, revolving around the planet with different velocities according to their respective distances."*

Saturn was the Solar System's last outpost, until the chance discovery of another world by a military deserter. Young William Herschel, a gifted musician, had answered the call of duty (and of his father) and joined the Hanoverian Guards as an oboist. Following a terrible defeat of the Hanoverian Guards by the French, young Herschel found himself lying in a ditch overnight. Perhaps he looked up at the stars and discovered his true interest. Whatever the reason, the 19-year-old deserted from the army and fled to England in 1757. Although being a talented musician, he was unable to afford the price of a telescope. Herschel was a man to overcome obstacles, and without any professional training, built himself a reflecting telescope.

In 1871, following his passion of searching for double stars, Herschel came across a greenish blob that looked initially like a comet: *"In the quartile near Zeta Tauri... is a curious either nebulous star or perhaps a comet..."* Despite an incredulous public stating: *"Who is this organist from Bath?"*, the musician had discovered a planet, later named Uranus. Herschel received a royal pardon from the British King from his desertion, became the official court astronomer and, together with his sister Caroline, pursued astronomy for the rest of his life. The planet itself, being far away, allowed little to be identified, though was inferred to be a gas giant. In the late 1800's, R.S. Ball wrote: *"Uranus also seems to be greatly swollen by clouds, in the same manner as are Jupiter and Saturn..."* However, by the 1900's Uranus was discovered to be tipped on its side, rolling about in its orbit like a rolling pin.

While trying to track the orbit of Uranus, astronomers found its motion was not quite in keeping with the new gravitational laws founded by Isaac Newton. The green planet seemed to jitter on its path around the sun, as if pulled by some mysterious force. This force could be explained if another planet was pulling on Uranus's orbit. A young Cambridge undergraduate, John Couch Adams, used mathematics to predict the location of this unknown planet. He wrote his solution in a letter that found its way to Astronomer Royal, George Airy. Dismissing Adams' word as that of an inexperienced youth, Airy threw away England's chance to discover Neptune. Instead, the Frenchman Urbain Le Verrier independently mathematically modelling the location of Neptune, received a message from Berlin University in 1846: *"The planet whose position you have pointed out actually exists."*

As with Uranus, the actual nature of the distant blue blob that was Neptune presented to the best telescopes was hard to determine. Herschel lamented: *"The discovery of Neptune is so recent, and its situation in the ecliptic at present so little favourable for seeing it with perfect distinctness, that nothing very positive can be stated as to its physical appearance."* Even a century on, former JPL Director Dr Bruce Murray stated: *"From here on out we know so little about the planets that we can hardly ask questions."* This view would remain more or less prevalent until Voyager 2's encounter in 1989.

The orbit of Neptune appeared to be pulled by the force of a mysterious planet X in a way that Uranus' orbit had. The possibility of yet another planet beyond Neptune intrigued many astronomers, including Mars studier Percival Lowell. Lowell in fact began searching for this mysterious planet from 1909 until his death in 1916 and had unknowingly photographed it twice. It wasn't until 1929 that the search for Planet X resumed, when Clyde Tombaugh, a young Kansas boy with no formal astronomical training, took up the task. On 15

February 1930, after months of painstakingly searching star plates using a blink comparator, Tombaugh ran to his supervisor, saying: *"Dr Langford I believe I have found your planet X."* Later named Pluto by schoolgirl Venetia Burney, the little world marked the end of the known Solar System before the space age.

The use of the telescope
for astronomy not only allowed humanity to understand the outer planets for the worlds they were, but also to discover 'solar systems' in miniature around those worlds. From the time of Galileo, it was known that there were moons around Jupiter and Saturn. As the industrial revolution allowed telescopes to become more sophisticated, the number of these moon discoveries increased, as tiny pinpricks of light were found around Uranus, Neptune and eventually Pluto. Although about thirty moons had been identified in the outer Solar System before the Space Age, almost nothing was known about them. Even the most powerful telescopes revealed them as nondescript, fuzzy discs. With little else to go on, astronomers used our Moon as a baseline, thinking that most of the outer Solar System moons would be colder, cratered versions of our own. Terrestrial snow-capped mountains and tundra provided additional inspiration, and scientists imagined worlds of icy mountains and craters, with geology being fairly similar between them.

An exception was Saturn's largest moon, Titan, where spectrometry earlier in the 20th Century had detected an atmosphere. Titan was envisioned to be the ideal space tourism destination, with views of the majestic ringed Saturn hanging in cobalt skies above an Arizona desert-like landscape sure to sell future tickets. As for the rest of the moons, little was written in textbooks before the Space Age, and their perceived blandness would relegate many of these worlds as a low priority for exploration with the coming of the Space Age.

The Space Age
for the exploration of the outer Solar System started in 1972. Until then, technology just was not up to the job of crossing the distances needed to reach the outer planets. Jupiter, the nearest gas giant to Earth, is five times more distant from the Sun than our world. Up until 1972, spacecraft sojourns had been restricted to the inner Solar System where trip times were measured in months. Journeys to the outer Solar System would be measured in years, stretching the reliability of electronics and systems available at the time to the utmost limit. Sunlight in this region of space was so weak that solar panels, the traditional source of spacecraft power, were considered useless. Radio signals transmitted to and from these far places needed to be much more powerful to be heard at either end, and would take hours for a round trip. Robot travellers were also on their own. There was no hope of in-flight repairs in case of failure, or the possibility of upgrades after launch. Moreover, the asteroid belt, full of probe-busting rocks, formed a formidable uncharted barrier between Mars and Jupiter.

Human factors also came into play. Scientists, living by 'publish or perish', had little desire to throw away parts of their careers designing and building a spacecraft, and then having to wait more years for a result to arrive from some far corner of the Solar System. That is assuming the spacecraft remained working, or got there at all. The idea of throwing away so much professional life only to have nothing to show for it seemed too great a risk.

Perhaps emboldened by the success of the Apollo missions, and with a spirit of 'we can do anything, we landed man on the Moon', NASA began planning for two spacecraft to make the first ever sojourn beyond the asteroid belt. In 1972, Pioneers 10 and 11 launched from Earth to try to pass the belt unscathed and reach Jupiter.

In February 1973, Pioneer 10 made it through to the other side of the asteroid belt and, finally, into the realm of the outer Solar System. As the spacecraft approached the giant planet, radiation levels started going up, and up, and up. The readings increased past the design limits and kept rising. False commands were being spuriously raised in the probe's fried electronics, as it received 1,000 times the lethal radiation dose for a human. At closest approach, the radiation levelled out and then finally decreased as the beat-up spacecraft hurriedly left Jupiter behind. Pioneer's crude images, occasionally marred by radiation damage, showed Jupiter's Great Red Spot to be an atmospheric storm, ending speculation of it being a landmass. Unfortunately, resolution was just too low to reveal much more about the Great Red Spot, or of the four Galilean satellites. These were discovered in the early 1600's by Galileo, and Pioneer's best results suggested

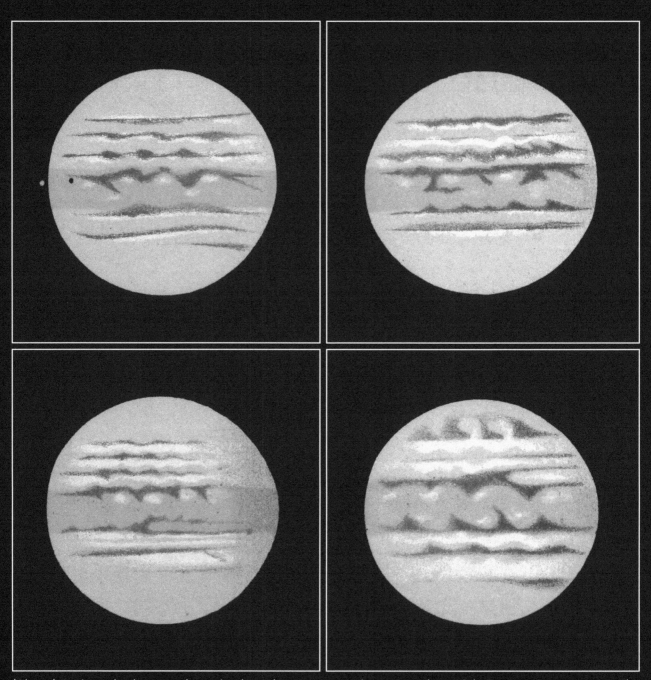

(Above) Jupiter, the largest planet in the solar system, as it appeared to 19th Century astronomers. The discovery of moons orbiting Jupiter by Galileo was stark evidence for a sun-centred, not Earth-centred solar system. Jupiter's fast 10 hour rotation and perpetual cloud belts were a mystery to early astronomers. Many thought the giant planet to be a younger, stormier version of Earth (Public Domain).

(Above) Saturn was the furthest planet of the solar system known to the ancients. This 19th Century artwork shows the majestic ring system that has inspired countless amateur astronomers. Saturn's rings perplexed astronomers; some thought they were solid, while others believed them (correctly) to consist of myriads of small moon-like particles, orbiting around the planet (Public Domain).

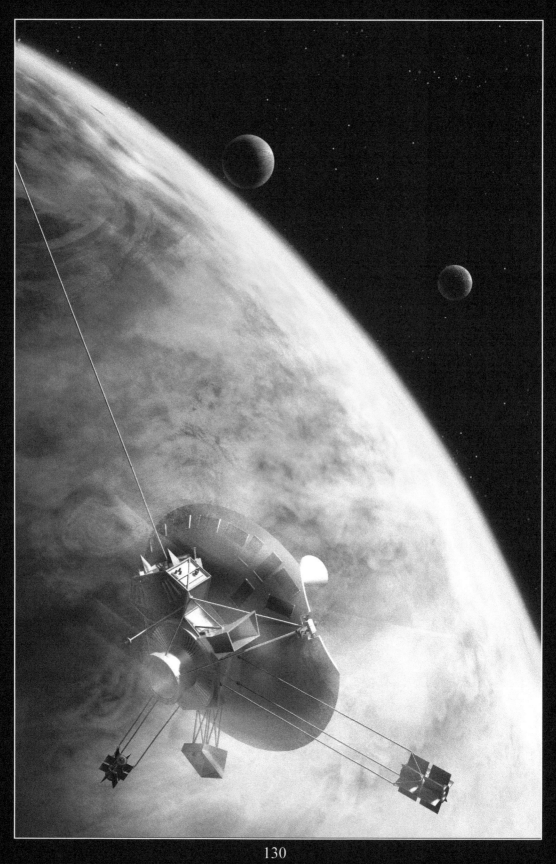

(Facing page) Pioneer 10, the first spacecraft to visit Jupiter, almost was fried by its massive radiation field. (Top) Pioneer imaged a view of Jupiter not possible from Earth as it flew near its pole (NASA). (Bottom) Pioneer 11 reached Saturn in 1979, almost smacking into one of its then unknown moons (NASA).

(Above) The two Voyager spacecraft, using far better imaging systems than Pioneer, radically changed humanity's perception of Jupiter and its Moons. Textbooks were practically rewritten overnight.

they were mostly made of water ice, and revealed tantalising markings on Europa and Io.

Pioneer 11 followed its sister craft past Jupiter, with a hastily modified trajectory to avoid the worst of the radiation. Humanity's knowledge of the Jovian system increased with Pioneer 11's encounter, as it added its results to those of Pioneer 10. With Jupiter receding into the distance from the speeding spacecraft, a unique celestial opportunity existed to send the little probe to Saturn.

Journeying further than any man-made object before it, Pioneer 11 reached Saturn in 1979. Although the first to reach the ringed giant, Pioneer 11's team was denied a close-up trajectory that would have returned the best possible data from the encounter. The NASA Director of the Planetary Program at the time, Tom Young, overruled his science team and instead used Pioneer 11 to trial a more distant path that would be used by the later Voyager missions, then under planning. This decision, while not popular at the time, made possible the once in a lifetime outer planet discovery mission of Voyager.

Despite this setback, Pioneer 11 discovered two more rings of Saturn and a new moon it nearly collided with on its way out into deep space. NASA Space Science Associate Administrator Thomas Mutch summarised the important part these little spacecraft played in space exploration: *"We have now moved out beyond the familiar part of the Solar System to explore planetary objects so unusual that their very existence*

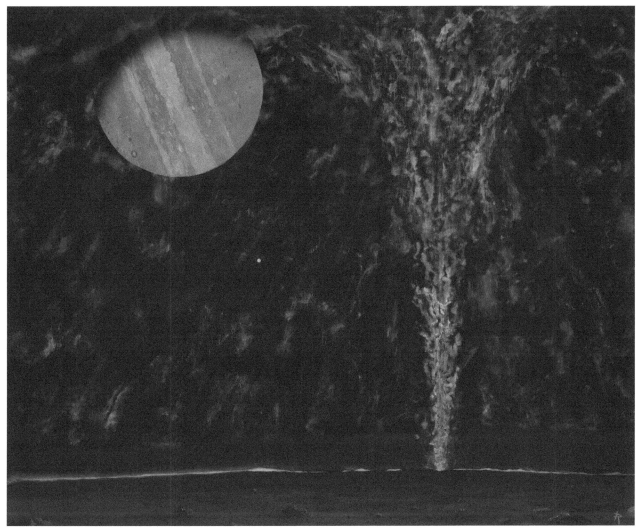

(Above) One of the profound discoveries of Voyager was that Io, one of Jupiter's moons is volcanic. Plumes of sulfur dioxide are shot far into the skies of Io (Jon Ramer).

was something people might accept intellectually but not really in an immediate sense."

The impact of the Voyager mission to society cannot be overstated. In the exploration of the outer Solar System, the impact of these two spacecraft on space exploration rivalled that of the Apollo Moon landings. Not only were astronomy textbooks made obsolete literally overnight, but two generations of engineers were inspired to design and fly four major follow-up missions.

"We all knew we were on a journey of discovery, but none of us imagined the wealth of discovery," said Ed Stone, Voyager Chief Scientist

In 1961, a graduate student discovered a once-in-176-year planetary alignment of the outer planets, which meant a spacecraft launched at Jupiter could use its gravity and that of Saturn and Uranus to slingshot itself all the way to Neptune. This way, almost all the outer Solar System could be explored in one fell swoop. The catch was that if 'The Grand Tour' missed its launch window between 1976 to 1978, NASA would have to wait nearly two centuries before trying again. Also, in order for all the outer planets to be visited, the spacecraft would have to last for at least 12 years. However, as former mission scientist Dr Tim Hogle stated, *"Missions in the 1960's were usually measured in days to weeks, and making spacecraft reliable for a 12 year journey to Neptune and Pluto was beyond our technical capabilities."*

The project also had to contend with massive budget cuts following the demise of the Apollo missions to the Moon and the lead up to the Space Shuttle program. Sufficient funding remained only to build two spacecraft guaranteed to last for four years — just enough time to reach Saturn. NASA was adamant that the probes would not be going to Uranus and Neptune, even assuming they could last that long, and the team should expect no changes to the flight plan that finished at Saturn.

Despite the NASA budget cuts, the two Voyagers were made state-of-the-art – and secretly designed with additional fuel to allow travel to Uranus and Neptune 'just in case'. The cutting-edge equipment created for the Voyagers brought a multitude of design problems that had to be fixed. Mission Designer Charley Kohlhase said: *"The mission design at that point needed to be redesigned... the Pioneer 10 and 11 missions had discovered that Jupiter's radiation levels were much higher than previously thought... [Project manager] Bud also wanted me to pick trajectories that would allow Voyager 2 to make a second pass at Saturn's moon Titan if Voyager 1 failed..."*

Finally, it was down to a choice of launching a faulty spacecraft or not launching at all, and Voyager 2 blasted off on 20 August 1977. As if to announce its disapproval at being launched early, the cantankerous spacecraft began refusing ground commands and firing its attitude control thrusters at random, and issuing other spurious commands to its scientific instruments. Just before arriving at Jupiter, Voyager 2's primary radio receiver failed and its sole backup developed the equivalent of tone deafness. However, with raw innovation that would mark the rest of the mission, engineers developed a workaround and saved the spacecraft. Voyager 1's launch on 5 September was almost uneventful in comparison and it overtook Voyager 2 to reach Jupiter in March 1979.

The effect the Voyagers had on the scientific understanding of the Jupiter system was nothing short of revolutionary. The trickle of data the Pioneer missions had sent from Jupiter several years before seemed to reinforce centuries of telescopic observations and theories. By contrast, the Voyager images flooded in on monitors, showing a planet full of psychedelic whirling eddies, hurricane-like jet streams and lightning bolts powerful enough to vaporise a city. Bill Kurth, Plasma Wave Science Co-Investigator wrote: *"Most of the time, these monitors displayed the most recent image from the Voyager Imaging Science Subsystem – the camera. The display was riveting. Each day and each hour, the images became clearer and more detailed, and new worlds were literally being unveiled before our eyes."*

Dr Tim Hogle summarised the situation the Voyager team faced: *"The sense of awe that we all experienced was completely mind boggling. To see pictures come in, one after another, each showing more detail than the previous one, did not allow us time to process what we were seeing before the next one arrived. It was so strange, we could only speculate."* The press had a field day during the encounter and the Voyager scientists did their best to explain to an eager crowd what they were seeing. Science on the fly would become the norm for subsequent Voyager encounters.

The surprises kept coming as the Voyagers started returning close-up images of Jupiter's moons. Convinced these worlds would be dull, many thought them hardly worth the trouble of wasting valuable mission time over, and it took the persuasion of Dr Carl Sagan and others to keep them on Voyager's list. Co-investigator Alan Cummings stated: *"I remember seeing the image of the moon Io for the first time and thinking that the Caltech students had engineered a brilliant stunt -- they must have substituted a picture of a poorly made pizza for the picture of Io! All that orange and black on Io changed our thinking about the moons in the solar system. I think most of us thought they would all look more or less like our own Moon. But, wow, how wrong was that!"*

Perhaps the greatest moon discovery occurred after most thought the picture show was over. A trajectory engineer by the name of Linda Morabito was performing routine checks on the orbit of Io, Jupiter's innermost Galilean moon. Viewing an enhanced image of the backlit moon, she saw what looked like a bright smudge on the edge of Io. The smudge turned out to be a sulphur plume, 280 kilometres high. Io had volcanoes! In fact, the little moon is the most volcanically active object in the Solar System, giving its surface a total facelift every thousand years. Small Europa proved almost as astounding. Its smooth surface likely

(Above) A montage, courtesy of Voyager, of Saturn and some of its moons. Titan is to the top right (NASA).

covered by a salty ocean that could possibly be an abode for life. Finally, as Jupiter receded behind them, the Voyagers returned one more surprise. Fine as cigarette smoke and only visible when backlit by the Sun, a ring was photographed around Jupiter.

The Voyagers' revelations of rings around Jupiter and other discoveries set a precedent for

what to expect during a close encounter with the outer Solar System planets. Expectations were high as the Voyagers sped towards Saturn – often seen as the jewel of the Solar System. However, the November 1980 Voyager 1 and August 1981 Voyager 2 encounters with Saturn exceeded everyone's expectations.

Strange spoke-like dark formations seem to float above the rings and Voyager 1 underwent a risky reprogramming to take a closer look. The rings themselves, originally thought to be homogenous bands of debris, revealed themselves to be made of hundreds upon hundreds of ringlets. An outer ring, the F ring, even turned out to be braided – a phenomenon later attributed to the gravitational interaction of shepherd moons. As with Voyager's Jupiter encounter, Saturn's moons dumbfounded mission scientists. Thanks to a massive prehistoric collision, the small moon Mimas sports a crater one-third the size of the moon itself. The collision almost smashed the moon to bits. Enceladus, as with Europa, suggested itself as an ocean world covered with ice. 'Yin and Yang' moon Iapetus could not seem to make up its mind whether it wanted a bright or dark surface, and took half each. The imaging team expected Voyager 1's look at Titan to top these successes and

135

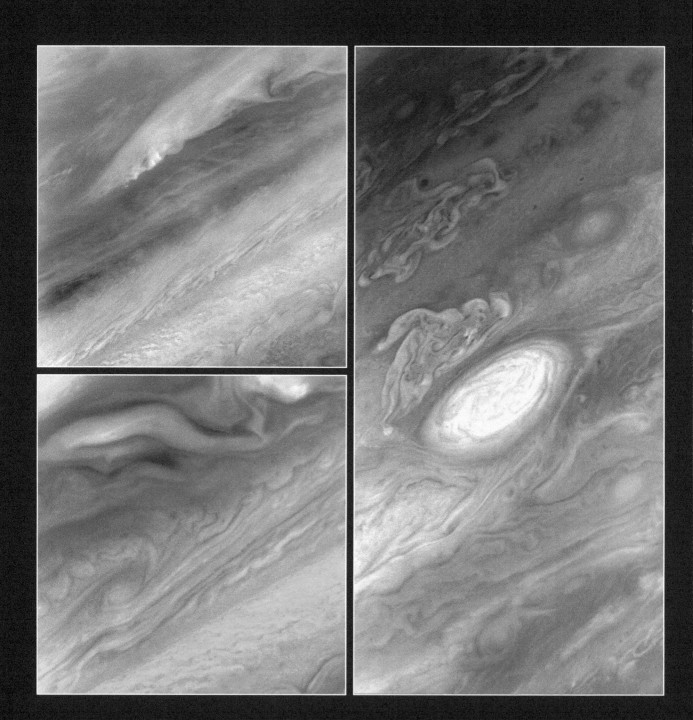

(Facing page) The swirling, almost psychedelic appearance of Jupiter's clouds imaged by Voyager were a surprise to scientists. Most thought that, this far from the Sun, Jupiter's weather would be more uniform (NASA). (This page) Voyager discovered the rings of Saturn to be amazingly complex, consisting of myriads of ringlets. These cast amazing shadows on the planet below (NASA).

(Above) As the Voyager spacecraft departed Saturn, humanity was treated to views of the planet and its rings not possible from Earth (NASA). (Facing page) The crescent Saturn hangs in the airless skies of Rhea, one of its moons.

be the crowning glory of the Saturnian encounter, but as with all the other surprises for the Voyager missions, Titan would be an unexpected puzzle.

The original planetary Grand Tour concept had planned for Voyager 1 to visit Jupiter, Saturn and Pluto. However, during mission planning the opportunity to view the only moon in the Solar System with an appreciable atmosphere close-up proved irresistible. Pluto was sacrificed, and Voyager 1 was reconfigured to make Titan its last flyby before leaving the Solar System forever. Titan began to grow in the monitors of Mission Control as Voyager 1 began to leave the Solar System. As Voyager 1 skimmed close to the orange moon's cloud tops, its cameras revealed… nothing. Titan's atmosphere proved too dense, hiding its surface permanently from view. The gamble had failed and the opportunity to explore Pluto was lost. The Titan encounter did return some data though, fixing the composition of its atmosphere and surface temperature.

Titan's surface would remain hidden until the Cassini Huygens mission dropped a probe on its surface in early 2005.

Voyager 2's Saturn encounter commenced in 1981 and provided closer looks at Saturn's atmosphere

(Above) A pre-space-age view of Voyager 2 encountering Uranus. At the time noone really knew what the planet, or its moons, would really look like (Michael Carroll).

and rings. Using its photopolarimeter — a device Jupiter's radiation had knocked out in Voyager 1 — Voyager 2 was able to provide pin-sharp imagery of Saturn's rings. Dr Linda Spilker used a photopolarimeter to reveal the complexity of the rings: *"One of my favorite memories involved watching the slow plotting of the occultations, the highest resolution Saturn ring data ever recorded... A pen moved back and forth carefully recording each data point on a long roll of paper that slowly unwound beneath it. I remember later carefully unrolling this scroll of data on the floor in a long hallway, looking at the amazing ring structure, and feeling like I was literally walking through Saturn's rings."*

After Saturn, many thought the Voyagers had done such a great job that no further discoveries were needed. Funding threatened to be cut again and Voyager 2, passing by Uranus and Neptune, would transmit to a world that was no longer listening. A stuck camera platform on Voyager 2, causing loss of some images partway through its Saturn encounter, didn't help matters. Fortunately, through hard work, the scan platform and threat of budget cuts were resolved, and Voyager 2 was able to continue with their 'just in case' plan to Uranus and Neptune.

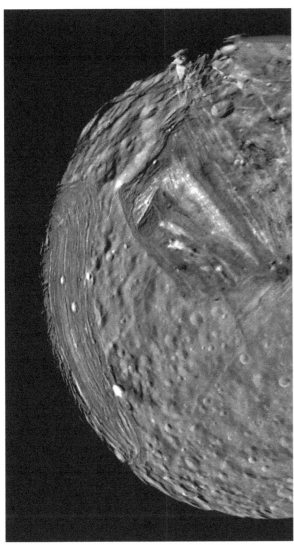

(Above) Voyager 2 showed Miranda to have extreme topography (NASA).

(Above) Uranus and its rings as seen from the bottom of the ice cliffs of Miranda.

As Saturn receded to a mere point of light indistinguishable from the background stars, the Voyager 2 team knew the spacecraft was literally entering uncharted territory. Voyager 2 was also travelling in a region it had not been designed for. Its 470-watt power supply at launch had decayed to just 400 watts. Its radio signal was also becoming ever weaker, almost lost against the background crackle of the galaxy. The all-important cameras, operating on a now fragile platform, would have to learn new tricks in order to operate at light levels only a quarter as bright at Saturn. Imaging methods used on the earlier encounters would no longer work in this Stygian darkness.

Dr Ed Stone explained: *"Because the Voyager spacecraft had been designed to operate out to the distance of Saturn, the 53-month gap between Saturn and Uranus was needed not only to respond to the scan platform problem, but to modify the spacecraft software to facilitate operation at twice the distance from the Sun and Earth."*

Voyager's imaging team realised they had to perform the longest distance system upgrade in history. Accessing the craft's critical flight operating system (a similar procedure that killed the Soviet Phobos 1 Mars probe in 1989), the programs for firing its manoeuvring thrusters were modified. Minute puffs would give Voyager enough stability to track its fast-moving targets, as well as keep its own movements in check. Voyager 1, far above the Solar System, was used as a flying test bed for its twin and was rather ingloriously experimented on for the sake of confirming the success of the new thruster program.

At last, four years after leaving Saturn, the green firefly-like globe of Uranus began to take shape on JPL's monitors. Voyager 2's encounter with Uranus was incredibly brief. Humanity's first close-up view of this distant world was crammed into a tiny six-hour encounter window, as Voyager 2 screamed past at over 18 kilometres a second. Again, mission scientists and the media packed the control room to receive the first close-up images of Uranus and its environs. As before, Uranus and its moons were surprising but not in the way that might have been expected. After being spoiled by the psychedelic storms and eddies of Jupiter and the rings of Saturn, Uranus proved to probably be the most un-photogenic planet in the known universe. A high level green atmospheric haze made Uranus as visually interesting as a tennis ball. Even high-end post mission contrast enhancement image stretching barely managed to discern the faint bands of the Uranian cloud system. Despite the rather lacklustre appearance of Uranus, engineer Sun K. Matsumoto was moved to share his experience: *"It was amazing to watch what everyone was doing and how it turned into these incredible images I had never seen before. I thought, 'Wow, we are changing science textbooks! We are making history!' It was such an infectious experience that I was seriously hooked."*

It was the moons of Uranus that were the highlight of the encounter. Pre-Space Age astronomy textbooks normally devoted less than half a chapter regarding the Uranian moons. Until Voyager 2, there was not much to tell; the five known moons appearing as dim points of light in even the best telescopes. Artwork of the time illustrated cratered icescapes, frozen rock-hard in temperatures that would instantly freeze-dry an unprotected human. As Voyager 2 closed in, Uranus' moon count was increased by 10, as tiny moons revealed themselves between the planet and its then known innermost moon, Miranda. Small Miranda was expected to be inert, and mission controllers almost half-heartedly awaited the first images of the little ice moon.

What appeared on the JPL monitors was a shock. Miranda looked like a patchwork quilt. Bits and pieces of radically different terrain made a hodgepodge surface, revealing evidence of at least five titanic impacts that had all but blown the moon apart. Arial, orbiting outside Miranda sported glacier laden valleys that seemed to originate from an early gravitational tug-of-war between it and the rest of the Uranian system. In fact, all the previously known Uranian moons except Umbriel showed evidence of early geological activity, involving large landscape shifts to create valleys, faults or slushy ice outflows.

Voyager's non-imaging teams received a surprise too. Expecting a magnetic field somewhere in the ballpark of massive Jupiter or Saturn, no field was detected around Uranus until the spacecraft was virtually on top of it. Uranus' magnetic field, like the planet, was also tipped over. John F. Cooper, Postdoctoral Scientist wrote of the magnetic field: *"We knew — of course — that Uranus was tipped over at 98 degrees to the plane of the Solar System, the ecliptic, but we had no inkling that the magnetic field of Uranus would turn out to be so strange, with a 60-degree tilt... It was totally amazing that our team was able to independently confirm the tilt by its effect on the absorption of radiation belt particles by the moons of Uranus."*

On 26 January 1986, Voyager 2 began to leave the green planet, preparing to enter the darkness once again. Even as it was photographing the crescent Uranus, tragedy struck at NASA that would leave the future of manned space travel in its own darkness. At the JPL press centre, media were struck by magnificent images of a receding green crescent on one television monitor, and the horrific Space Shuttle Challenger explosion on another. Challenger affected later space missions, such as the Venus Magellan probe, but in a terrible moment for Voyager, the Uranus discoveries were pushed into the background. Ironically, Voyager 2's triumph at Uranus and journey to Neptune almost became the only activity of significance in western space exploration. The entire Space Shuttle fleet, and the will to launch new space missions, became grounded pending a thorough investigation. John F. Cooper recalled the time: *"The five days of January 24 to 28, 1986 — starting when Voyager 2 flew by Uranus ('the planet that got knocked on its side') and ending with the painful tragedy of the Challenger accident — are forever etched in my memory of unforgettable life experiences."*

Voyager 2 continued to cross oceans of space on its way to Neptune — the furthest planet in the Solar System. The planet itself is so far away that it hadn't even been discovered until 1846. Neptune reflected the

(Above) As Voyager 2 sped past Neptune, its last planetary encounter, it looked back at the crescent planet and largest moon, Triton. Voyager's mission at Neptune was both a first look and final farewell. Both Voyager

almost 4.5 billion kilometre distant Sun's light so faintly that the best telescopes in the world at the time of Voyager revealed it only as an indistinct blue ball. At Neptune's distance, Voyager 2's signals travelling at the speed of light would take four hours and six minutes to reach Earth, and the spacecraft would arrive in August 1989, 12 years after it had launched.

Time had taken its toll on the solitary explorer. As if afflicted by worsening Alzheimer's disease, an increasing amount of Voyager 2's processor memory had succumbed to the harshness of space. Dwindling energy supplies would also have to be rationed, and collection of scientific data needed to be prioritised in order not to overload the frail spacecraft.

Time had also changed Voyager's control team on Earth. Typewriters used by scientists and the media alike at JPL for the Jupiter encounter had been replaced by computer laptops. America, Japan, Spain and Australia were also pooling their resources to catch the ten quadrillionth of a watt signal of Voyager 2 as it closed on its distant and final planetary target. Even scientists of the former Soviet Union were participating in science operations. This was a gesture of international cooperation that was a sign of the thawing of Cold War tensions between the two world superpowers.

Coaxed by its controllers for one final push, Voyager 2 arrived at Neptune six minutes late on 24 August 1989 and flew within 5,000 kilometres of the planet's north pole. On the way in, it had already discovered a number of small moons, one of which was larger than Nereid, the smaller of Neptune's two known satellites. Neptune's 'ring arc' system had also been investigated. Unlike the other three gas giants, Neptune's thin rings were clumped in places, looking like incomplete bands circling the blue planet. Proposals of yet to be discovered shepherd moons, or even a recent impact between a comet and a small moon were announced to try and explain the mysterious ring arcs, but no-one was certain of the correct explanation.

General Science Team member Trina Ray recalled Voyager at Neptune: *"As encounter day approached, there were press camped out in trailers down Mariner Road for as far as the eye could see. Every day, the smudge of Neptune kept getting bigger and bigger and more resolved and more beautiful on the screens showing the latest images. And — there were screens everywhere! All around the lab, in the cafeterias, in the conference rooms, in the hallways — you couldn't help but be caught up in the ever-growing resolution of the images."*

Some of the greatest surprises came from Neptune itself. In a final defiance to established planetary science, Voyager 2 rewrote the text books as it skimmed over the planet's cloud tops. For decades astronomers taught that Neptune would almost be a dead world, its half-frozen atmosphere barely stirring in the Stygian cold. The Sun simply did not provide enough energy to power significant weather systems in a world this far out. As Voyager 2 approached, distinct markings resolved themselves on Neptune's disc. A huge dark spot, echoing the Great Red Spot on Jupiter, moved at an astonishing speed against Neptune's direction of rotation. Small high altitude white clouds formed and vanished in a matter of minutes, screaming past the dark spot at over 2,500 kilometres per hour. Voyager revealed that Neptune possessed the fastest atmospheric winds of any Solar System planet, flying in the face of established doctrine. Two hours before Voyager 2 made its closest approach to Neptune, it photographed a bank of cirrus clouds, casting long shadows into Neptune's main cloud deck, 100 kilometres below.

Skimming past the blue planet's cloud tops, Voyager 2 made its approach to its last Solar System target, Triton. At nearly four in the morning, scientists sustained by junk food fixes watched in awe as the first images of Neptune's largest moon filled their monitors. Triton's cold surface, the coldest in the Solar System, was a mixture of craters, frozen lakes of nitrogen, and landscapes subjected to local heating events. Triton's greatest surprise appeared at its south pole. As Voyager photographed that area, imaging team member Laurence A. Soderblom noticed a series of dark smudges on the slowly evaporating polar nitrogen ice. Could they be formed by geysers, he wondered? Incredulous as it was, later images confirmed Soderblom's theory. Triton, the coldest body in the Solar System, was volcanically active. Heated by the far away Sun, pockets of vaporised nitrogen built up enough pressure to break free of the surface and erupt in a thin jet. Revealing Triton's unexpected volcanism was a final and fitting tribute to Voyager's discoveries. Dr Ed Stone recalled: *"The Triton flyby was my favorite moment partly because it was a bookend."*

The image of crescent Neptune shrank on JPL monitors, gradually merging with the stars. More than

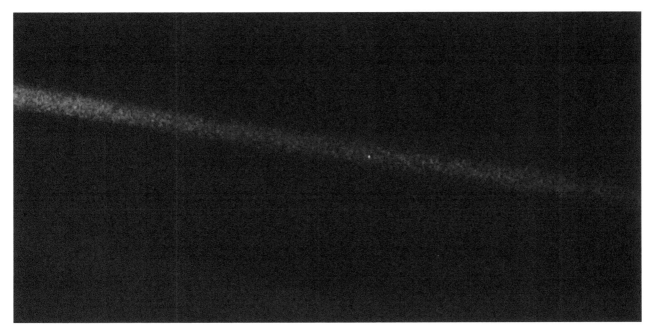

(Above) In 1990 Voyager 1 took the most distant photograph of Earth ever taken; a pale blue dot suspended on a mote of dust (NASA).

a few members present in JPL's control room didn't want to let the little probe go. As Chuck Berry's 'Johnny B. Goode' played — a song included on the gold record attached to the spacecraft Voyager — to a farewell gala party at JPL, Voyager 2's cameras were commanded to switch off. The eyes that had rewritten textbooks of the outer Solar System closed forever. However, this was not to be the last the public would hear from the hardy little spacecraft, and Voyager 2's twin would soon be preparing for one final picture show.

By 1990, Voyager 1 had virtually been forgotten by the public at large as it careened out of the plane of the Solar System, alone. Its iconic Saturn and Titan images were already eight years old and there seemed nothing left for it to photograph, billions of kilometres from the nearest planetary body. However, a small group of scientists led by the television series Cosmos presenter Carl Sagan, realised Voyager 1's position would provide a once-in-a-lifetime opportunity to image the Solar System from the edge of space. The proposal met with strong resistance; the risk to the spacecraft was too high and there was no scientific value to the images. Sagan's personality and influence won through, and the aging spacecraft's imaging system was gingerly reactivated.

On Valentine's Day, 1990, Voyager 1 took 61 photographs of the Solar System, seen from almost six billion kilometres away. The distance proved too great for Mercury, Mars and Pluto, but one of the remaining planets was captured, a small blue world shining through the glare of its parent star — the Earth. This was the pale blue dot, the home to all humanity. The Voyagers had come full circle. Their images had rewritten textbooks and revealed a Solar System that was more fantastic than had been imagined. In the final picture show, however, as in the Apollo missions before them, the cameras were turned home, revealing a beautiful, fragile planet suspended in space. This image is one of the iconic views of the Space Age. During the time the Solar System mosaic was displayed at JPL, the picture of Earth had to be replaced often, as people would continually touch it as they passed by.

Somewhere in JPL, a computer terminal marked 'Voyager Critical Hardware' marks the current monitoring station for both of the Voyager spacecraft. Clever rationing of the Voyager's power supplies and dwindling propellant stores have allowed them to survive long enough to cross the heliosheath, the last remnant

(Above) En route to Jupiter, the Galileo spacecraft helped investigate the collision of Comet Shoemaker-Levy 9 into the giant planet (Mark Pastana).

of the Solar wind. Voyager 1 reached interstellar space in 2012, finally leaving the Sun's sphere of influence. The Voyagers have even survived the threat of budget cuts, the latest of which was in early 2008. However, lobbying by volunteer interest group The Planetary Society and other concerned members have extended their operations. At the time of writing both Voyagers have survived over 40 years since they had been launched. Although most of the Voyager team have moved on, some working on the later Galileo and Cassini missions, those who worked on the Voyager missions agree that it was the journey of a lifetime. Attenuated by distance, the now extremely faint signals from the Voyagers take many hours to reach Earth. Decades on, Voyager's images still provide an iconic view of our neighbouring worlds, and a monument commemorating their epic journey can be found at the Deep Space Network tracking station at Tidbinbilla, Australia.

Voyager's profound discoveries of the outer Solar System left many wanting more, particularly of
Jupiter. The limitations of trying to cram as much science as possible into a few short hours of a flyby were sorely felt, and the giant of all planets certainly warranted a more investigative mission. Unlike diminutive Voyager, such a spacecraft would need to be huge to carry the fuel needed to enter Jupiter's orbit. Huge was bad as it made it expensive to build, launch and fly. Miraculously, such a spacecraft, named Galileo after the discoverer of Jupiter's larger moons, was built, though had the misfortune to be prepared during a time of upheaval for NASA. First were the inevitable setbacks from trying to build a spacecraft of that size.

Compromises were forced on the design, including a foldable main antenna to make it more compact for launch. One NASA engineer cautioned: *"If the high gain antenna fails to deploy, we're dead in the water."* Secondly, as Galileo was being finalised, the Space Shuttle Challenger disaster struck, cancelling future manned flights and sending the space agency into a quandary for three years. Politics of the time precluded using a rocket to launch the spacecraft and Galileo was forced to wait, along with the rest of NASA, on the outcome of the Challenger review. During this time, Galileo's plutonium power source, prepared and installed for a planned earlier launch, became partially depleted. The spacecraft was also shot at, likely by protesters, during one of its many road trips between JPL and Cape Kennedy.

Finally, on 18 October 1989, astronaut Shannon Lucid sent the already aging spacecraft on its way from Space Shuttle Atlantis. Over the next six years, Galileo looped around the inner Solar System in a

(Below) This magnificent image of Jupiter by Juno was built on the legacy left from its predecessor, Galileo (NASA).

compromised trajectory forced upon it by the decision to launch using the Shuttle. The probe endured more heat than it was designed to as it built up speed to reach Jupiter, then disaster struck. Galileo team member Erik Nilsen explained: *"... shortly after the first Earth flyby, the operations team at JPL commanded the [high gain antenna] to open. After twenty minutes of anxiously waiting for the fully deployed signal, the project team realized that something terribly wrong had occurred..."*

No high gain antenna meant no Jupiter images. Over the next two years, engineers kept trying to deploy the antenna, while a separate team worked desperately to invent novel data compression techniques, and upgrade listening stations back home. Galileo was able to conduct the first ever close flyby of asteroids en route, even discovering a tiny moon orbiting one of them.

Eventually, on 7 December 1995, Galileo entered Jupiter orbit. On the way it had deployed a sub-probe that slammed into the Jovian atmosphere with all the grace of a meteor. Surviving less than an hour before being crushed and fried, the sub-probe returned valuable information about Jupiter's atmosphere. NASA wrote in a statement: *"It appeared that Jupiter's atmosphere is drier than we thought. Measurements from the probe showed few clouds, and lightning only in the distance. It was only later that we discovered that the probe had entered an area called a 'hot spot.'"*

Over the next eight years, the hobbled Galileo, trickling images and science data through a less powerful low gain antenna, scrutinised the Jovian system, making profound discoveries that Voyager could not. Firstly, unlike Voyager, Galileo's cameras were able to see the colour red, revealing Jupiter and its moons in true colour for the first time. Europa, with its salty ocean, was also high on the science list. Covered in what looked like arctic pack ice, parts of Europa's ocean squirted out into space from tidal action, leaving a briny deposit behind. Could these oceans contain life? As with all searches for extra-terrestrial life so far, Galileo could only go part way to answering that question. Europa's mysteries would have to wait for a future mission. In the meantime, Galileo's sensitive instruments surprised scientists by detecting evidence of oceans inside Ganymede, and the supposedly geologically dormant Callisto. Io was too far inside Jupiter's lethal radiation for Galileo to investigate too closely, however, cautious glances from afar identified dozens of new volcanic vents. Almost re-inventing itself since the Voyager flybys, Io truly was a moon turning inside out.

Galileo's mission was extended to the 2000's

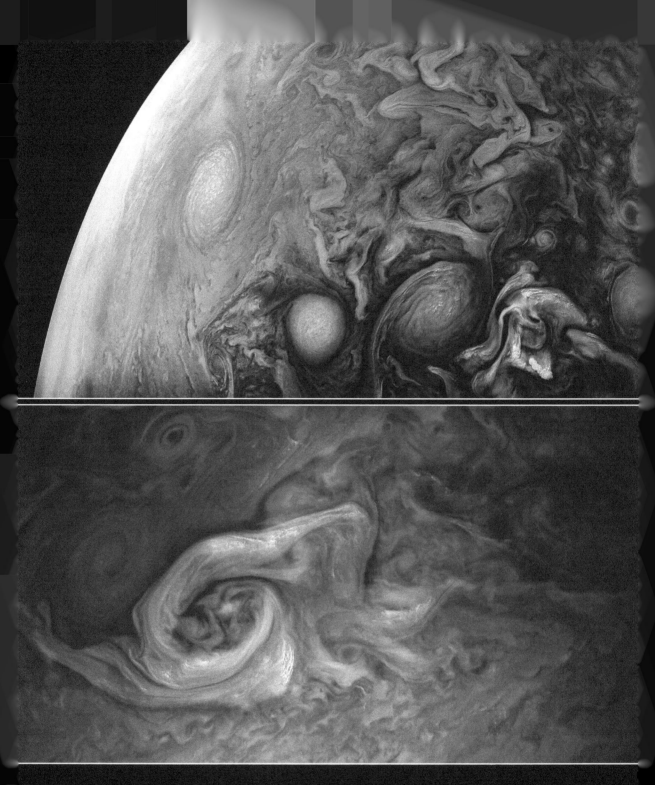

(These pages) The Juno spacecraft's primary mission at Jupiter was not to take images. In fact, the camera was primarily included to increase public engagement and encourage citizen science. Nevertheless, Junocam's images reveal the complexity of Jupiter's turbulent weather in unprecedented detail. Scientists are now able to better study the interaction between different weather systems such as spots and clouds on Jupiter (NASA)

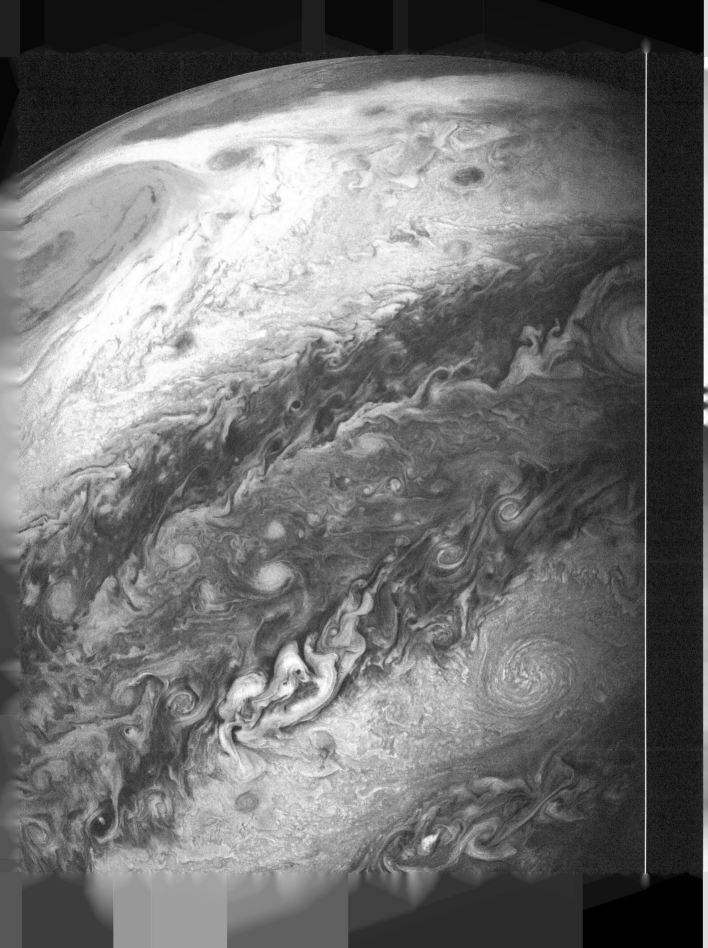

(Above) The Cassini spacecraft, carrying the Huygens sub-probe commences its perilous journey through Saturn's ring system. Cassini is about to fire its rocket motor to enter orbit around Saturn.

but radiation and dwindling fuel was taking its toll on the spacecraft. On 14 April 2003, in order to prevent Earthly bacteria potentially contaminating life on Europa, Galileo followed the path of its sub-probe into Jupiter's atmosphere, burning up and disintegrating.

As is usual in science, Galileo raised almost as many scientific questions about Jupiter than it answered, and a follow-up mission, Juno, was soon planned. So why return to Jupiter? Mission team member Lucyna Kedziora-Chudczer explained: *"The mission is looking at the questions of origins of the solar systems also related to understanding [Jupiter's] interior… We can get very deep mapping which wasn't possible with Galileo."*

A world-wide shortage of plutonium forced NASA to rethink the power design for the mission. Juno would instead be the first to carry solar panels that far from the Sun. The panels required, and the spacecraft to carry them, would be massive. Tracy Drain, Deputy Chief Engineer stated: *"So, we need to have a lot of area on our solar arrays because the spacecraft is so far away from the sun. And also over the last 20 years, we've had about a 50-percent increase in solar array efficiency. And that makes it easier to do at that distance."*

To complete its mission, Juno would have to venture closer to Jupiter than any mission before it, in the full force of Jupiter's deadly radiation belts. Principal Investigator Scott Bolton outlined the design: *"Juno is basically an armoured tank going to Jupiter. Without its protective shield, or radiation vault, Juno's brain would get fried on the very first pass near Jupiter… The background radiation that we're exposed to on Earth is about a third of a Rad. What we expect to see at Jupiter is about 20 million Rad."*

Another first for Juno was its science payload. For the first time, the camera system on board Juno would not be the primary focus of the mission; other instruments would do the heavy science. The camera was

(Above) Cassini flies through geyser plumes of Enceladus, one of Saturn's moons. Like Jupiter's moon Europa, the surprisingly active Enceladus may hide a subsurface ocean and, potentially, life.

instead deliberately designed for public relations, where raw imagery would be offered up for exploitation by amateur scientists and enthusiasts. Also for the first time, musician Vangelis, who composed the score for the film 'Chariots of Fire', was approached to create a special soundtrack for the mission.

On 5 August 2011, Juno blasted off from Cape Canaveral and spent five years swinging past Mars and Earth to steal enough speed to reach Jupiter. In July 2016, Juno fired its main motor, entering Jovian orbit with the following words from Mission Control: *"Juno, welcome to Jupiter"*.

Following a polar orbit and skimming within 8,000 kilometres of the cloud tops, Juno returned breathtaking views of the planet's poles and Great Red Spot. These were long awaited images that were denied the short-changed Galileo spacecraft, unable to transmit such high data volumes. Juno had just one year to complete its mission before radiation gave it the robotic form of terminal cancer. After that time, Juno would follow Galileo before it, plunging beneath Jupiter's clouds to burn up and disappear forever.

On 15 September 2017, a meteor ploughed into Saturn's atmosphere, glowing increasingly brighter over the period of a minute. Finally, the fireball violently exploded with the force of a one kiloton nuclear blast, before all fell silent once more in the eternal Saturnian skies. The explosion marked the dramatic end of the Cassini mission's 13-year exploration of Saturn and its moons, and nearly 30 years of planning, building and journey through space. Cassini Program Manager Bob Mitchell wrote: *"We're going to Saturn...a question that I hear is... What do we expect to learn? ... We really don't know. That's why we're going."*

Cassini was born from the legacy of the Voyagers. Many scientists had hoped that these spacecraft would solve the mysteries around this famous ringed planet, however, the missions had only raised as many

(These pages) Cassini was finally able to succeed where Voyager failed; lifting the veil on Titan. The only moon in the solar system with an appreciable atmosphere was a spectacular target for Cassini including (above) being imaged through the plumes of Enceladus (NASA/Don Davis).

(Above) Cassini imaged Saturn's rings in unprecedented detail, here shown backlit by the sun. Saturn itself casts an immense shadow on the right portion of the rings (NASA).

new problems as had been solved. Saturn's rings were far more complex than previously thought, complete with braids more at home in a jewellery shop, and mysterious spokes that seemed to defy physics – *"It's not something that just clicks into place,"* stated mission scientist Jeffrey Cuzzi during the Voyager 1 encounter.

Then there was Titan, the only moon in the Solar System with a significant atmosphere. A Pluto flyby was sacrificed in order for Voyager 1 to try and look at that surface hidden by clouds of hydrocarbon. Although valuable information was gained from Voyager 1's heat analysing instrument and radio science team, not so much as a glimpse was obtained of Titan's surface. Linda Spilker outlined the situation: *"From Voyager, Titan was so unusual. All we could see from Voyager is this hazy world. We couldn't see through to the surface. And that's why we had to go back with Cassini carrying the Huygens probe to try to see what the surface of Titan looks [sic] like."*

A new mission would return to the Saturnian system for a closer look. In 1990, plans were put in place to devise the largest and most complex spacecraft that had ever been sent to another planet. The undertaking was so ambitious that the Cassini Huygens mission needed two space agencies and 14 years to plan and build. The Cassini spacecraft was built by NASA while the Huygens probe was the responsibility of the European Space Agency (ESA).

One of the major hurdles to overcome was the Titan landing. No one had more than a basic understanding what the environment of Titan was like. Its hydrocarbon atmosphere proved an opaque veil to Voyager's cameras and scientists weren't even sure whether the ground beneath was solid or liquid. Huygens would have to be prepared for both. A second hurdle was budget cutting. Cassini team member Bob West

(Below) In 2005 the Huygens probe made the most distant soft landing ever attempted when it touched down on the surface of Titan (Ray Cassel).

Our understanding of Jupiter has changed massively since early telescopic sketches of the giant planet (top left). Pioneer's view of the Great Red Spot started providing hard answers on the atmosphere of Jupiter (Top Right). Voyager rewrote the textbooks on Jupiter (bottom left) while Juno (bottom right) presented a new generation of scientists and enthusiasts unprecedented views of Jupiter (Public Domain, NASA).

Similarly, the space age shaped our understanding of Saturn and its rings. From early telescope sketches (top left) through Pioneer 11 (top right), Voyager left scientists spellbound by discovering the complexity of the rings (bottom left) while Cassini has left a legacy of amazing images (bottom right). One of the many discoveries from Cassini was a bluish haze above the mid-latitudes caused by seasonal ring shadowing (Public Domain, NASA).

explained: *"Cassini underwent a major cost-cutting exercise early in the design. As a result the planned scan platform was tossed. There were many consequences of losing the scan platform. First, now the entire spacecraft must turn in order to point an instrument. That's not so bad by itself, but now there are a lot more conflicts to work out to try to satisfy the pointing needs of the different instruments."*

The Huygens probe mission was also trimmed, with a survivable Titan landing a 'nice to have' rather than a design requirement. Engineers pressed on, though, and did their best to make the Huygens probe as robust as possible to deliver the first ever surface pictures of an outer Solar System moon. In a manoeuvre tried by spacecraft before it, Cassini would use the gravity assist of three planets to slingshot its way to Saturn. Venus and Earth flybys after launch would send it to Jupiter, whose huge mass would finally fling it to Saturn after a seven-year journey. A rocket engine would then fire as Cassini dove through Saturn's rings, and everything would have to work perfectly. The 90-minute time delay between Earth and Saturn meant that Cassini would be on its own. As such, the spacecraft would need the smartest processor system ever flown to space, using critical sensor information to autonomously control its entry to Saturn orbit. Along with shrinking budgets, the passage of time also brought advances in micro-electronics. The quirky and fragile vidicon cameras of the Voyagers were replaced by state-of-the-art charge-coupled device (CCD) digital cameras; the measly 4 kilobytes of memory Voyager had to contend with was upgraded by factors of hundreds.

As Cassini took shape, one engineer in particular was taken aback at the sheer size of their design: *"It's just a monumental machine. It's the individual people that all put their pride in putting this together and building it right."* The conjoined probes weighed nearly six tonnes and were the size of a school bus.

Prior to launch, one more hurdle needed to be overcome: Cassini needed nuclear power to survive in the dimness of Saturn where solar panels would be useless. A radical group tried desperately to ground the mission on fears the plutonium would pollute space. Although trying court injunctions and even breaking into the grounds of the launch facility, their efforts failed, and Cassini lit up the Florida skies on 15 October 1997 with the words from Mission Control: *"3-2-1- and lift-off of the Cassini spacecraft on a billion-mile trek to Saturn!"*

The years following launch kept the Cassini team busy. Bob West explained: *"... there was a lot to do for the science teams to plan the mission — we had to decide how many orbits, what the orbit parameters should be... We also had to split up the time among the many, many competing requests for priority to look at various objects."* In 2000, as Cassini flew past Jupiter, it gave a unique opportunity to conjoin its investigations with Galileo, in orbit around the planet. In an unprecedented activity, two spacecraft simultaneously investigated a gas giant.

On 1 July 2004, at the end of a seven year journey, Cassini prepared to enter Saturn's orbit. At 90 light minutes away, Mission Control could do nothing to help with this critical manoeuvre. Cassini was on its own. Bob West: *"During the orbit insertion my main concern was that the thrusters fire with the right thrust and duration and spacecraft attitude to put us into orbit. There were two Mars missions lost at this critical juncture and so for me this was a little bit of white knuckle time."*

Cassini's radio signal, having disappeared as it flew behind the planet during the burn, re-appeared and was met with applause and cheers from Mission Control. Cassini had arrived! Even as the first Cassini images of Saturn were received by JPL – the first close-ups since Voyager — it was clear that Saturn was not the planet that Voyager had revealed. For starters, the northern hemisphere of a normally beige Saturn was blue. This half of Saturn, fully lit during Voyager, had been shaded by the rings by the time Cassini arrived, causing a type of smog that shrouded part of the planet to clear. No one had expected this; not even images from Hubble had shown it. As for the rings themselves, first looks at images returned from Cassini were often encounters with a few seconds of wordless silence. Cassini team member Linda Spilker explained: *"We turned the Cassini cameras down to look at the rings, revealing them in a way we had never seen them before. I remember coming back to JPL early in the morning just so I could be there and watch those pictures one by one come down. And I felt like I could almost reach out and touch the rings that were right there."*

Images of the stupendous complexity of the rings, of rainbows dancing between the particles, sunrises and sunsets of the Saturnian system, and multiple lunar eclipses played out to worldwide audiences for the whole Cassini mission.

Around Christmas 2004, Cassini released the Huygens probe it had carried for so long for the highlight of the mission — a landing on Titan. Bob West: *"The biggest nail-biter was the Huygens descent. Things had to work autonomously and correctly under new and trying circumstances."* Huygen's heat shield reached twice the temperature of the Sun's surface as it screamed into Titan's atmosphere. A series of parachutes slowed the craft and a camera began revealing the surface Voyager 1 failed to see. Also, unlike Voyager, the Huygens landing was streamed live via the internet. Although the images would take some time to transmit across 1.6 billion kilometres of space, audiences around the world could follow the probe's progress.

As the smog cleared, stunningly Earth-like images appeared. Linda Spilker: *"[The surface] had river channels. It had lakes… The only difference is, it's very cold on Titan. And the liquid that flows through Titan's rivers is methane instead of water."* Under the eternal ochre twilight, methane, normally a gas on Earth, would rain down in liquid form onto the icy surface, with drops as big as tennis balls. Flash flooding would gouge out deep channels in the normally dry surface, forming temporary rivers and carving the landscape that Huygens saw.

The probe landed with the grace of someone jumping out of a first storey window into a muddy paddock. Slowly freezing in minus 200-degree celcius temperatures, Huygens transmitted images of a stream bed replete with rounded ice stones for 10 hours before its batteries finally gave out. Dr Earl Maize of JPL was on hand as the Huygens data was received: *"It's just exhilarating. There were boulders. There were pebbles. There's a dry lake bed. And I still get goose bumps just talking about it."*

Later discoveries by Cassini revealed methane lakes at Titan's poles, and possible cryovolcanism where ice and methane took the place of terrestrial rocky lava. Pre-Space Age artists had depicted Titan as a beautiful world, with the best views of the Solar System. The ringed Saturn, hanging in the cobalt blue skies of Titan, where craggy spires reached up into the heavens, had inspired a generation of astronomers. Instead, in a somewhat recurring theme, the Space Age had revealed Saturn's largest moon to be even more surprising than artists had envisioned. Instead of a blue atmosphere, the heavy smog delivers once-in-a-millennia monsoon, where oversize droplets descend in slow motion to erode dunes and hills made of hydrocarbon sand. Cassini's Titan discoveries would keep researchers busy for years to come.

Cassini's mission was so successful that it was extended twice, and for 13 years the spacecraft changed hamity's understanding of Saturn and its Moons. Geysers were found on Enceladus, revealing a subsurface ocean, a billion-and-a-half kilometres from Earth. In almost spur of the moment planning, Cassini was reprogrammed to fly through them, barnstorming the wisps of water vapour and tasting them with its instruments. Perhaps Cassini's greatest gift to humanity, though, were the pictures. Through its mission, Cassini captured thousands of images that rivalled the best artwork ever created. JPL's Shadan Ardalan stated: *"As an avid photographer, these images for me contain as much artistic value as they do scientific value."*

The Cassini mission was again extended until in 2016 a critical decision had to be made. The spacecraft's fuel was almost gone (one percent with a two percent error margin). In order to prevent accidental contamination of Enceladus' potential living oceans, Cassini flew by Titan one last time to perform high risk loops between Saturn and the rings. Finally, on 15 September 2017, Cassini took a suicidal dive into Saturn's atmosphere. With thrusters strained to keep its antenna pointed towards Earth, and returning data up to the last second, Cassini finally broke up and became part of the planet it had explored for so long. Earl Maize was on the flight controller loop at the time: *"We just heard the signal from the spacecraft is gone and within the next 45 seconds so will be the space craft. I hope you're all as deeply proud of this amazing accomplishment. Congratulations to you all. This has been an incredible mission, an incredible space craft, and you're all an incredible team. I'm going to call this the end of mission. Project manager off the net."*

In the skies of the far reaches of the Solar System, the brightness of high noon is equivalent to

a moonlit night on Earth. On 14 July 2015, a small golden speck glinted briefly as it streaked through the cold sky, before disappearing from sight. The brief encounter marked the highlight of the New Horizons mission to Pluto that was 15 years in the making.

Until 2015, distant Pluto remained the last world of the original Solar System yet to be explored in detail. Circling almost 4.5 billion kilometres from the Sun, the small world was only discovered in 1930

(Above) New Horizons was able to perform the last rewrite of the original solar system by visiting Pluto and Charon in 2015. Both worlds proved to be surprisingly complex, with diverse geology (NASA).

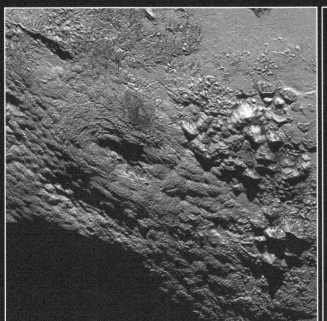

(Top left) New Horizons imaged Wright Mons, a possible ice volcano on Pluto. (Top right) Looking back at the Sun, Pluto's atmosphere was shown to be blue in colour, similar to Earth. (Above) As New Horizons made its closest approach to Pluto, it saw mountains surrounding a vast plain made of nitrogen ice. The likely once in a lifetime encounter also discovered crawling glaciers, again made of nitrogen (NASA).

by Clyde Tombaugh. Even with the Hubble Space Telescope, the best maps of Pluto were worse than what pre-telescope astronomers saw of the Moon. In terms of size, Pluto wasn't much bigger than our Moon. This might have made it unworthy of exploration compared to closer targets such as Mars or the larger gas giants. The bitterly cold world with an assumed endless changelessness of geology did little to attract commitments of a fragile spacecraft whose travel time would be measured in decades. Conversely, other researchers argued, the pristine nature of Pluto should be exactly the reason for a closeup look. Investigation of Pluto's surface could provide clues as to the birth of the Solar System, along with details on what bodies are made of so far out in space.

One researcher particularly keen on Pluto was Dr Alan Stern. Studying Pluto in the 1980's, he discovered the possibility that Pluto might have a part-time atmosphere. Being so cold and with an elongated orbit, most or perhaps all of Pluto's atmosphere might freeze onto the surface as the planet moved away from the Sun. Clearly Pluto was worth a closer look. Alan Stern demonstrated to an audience how hard it was to study Pluto from the ground by using an image of Earth degraded to Pluto standards: *"What could you learn about the earth at that resolution? Not very much, you can't even tell that there are continents and oceans on the planet. This is the reason we send spacecraft far away is to get close so we can do real science."*

A flyby from Voyager 1 was sacrificed in order to vainly try to penetrate Titan's smog. Many proposals were offered up, however, were continually refused funding in favour of a virtual gravy train of Mars missions. The target was always seen as too small and too far away for a closer look. Eventually in 2001, and probably due to pressure from school groups mad for Pluto, Stern received funding for a Pluto mission that had to leave Earth by 2005. Facing scorn from his peers, and the pressure of building with a budget a fifth of the cost of the Voyagers, Stern virtually pulled off a miracle. In 2005, the piano-sized New Horizons spacecraft was launched from a rocket meant for far heavier payloads. The result: New Horizons screamed past the Moon's orbit in just nine hours (the Apollo spacecraft took three days) and became the fastest artificial object to date.

Nine years later, New Horizons, unable to slow or stop, flew past the Pluto system and for the next 15 months painstakingly transmitted images at data speeds that were considered slow in the 1970s. Until the flyby, no one really knew what Pluto was like. The large moon Charon was discovered in 1978 and another four moons discovered in the 2000's but, apart from being reddish, the surface was a mystery. New Horizons team member John Spencer explained: *"I'm feeling pretty exhilarated at this point. You're at the top of the roller coaster. You're about to go down that dizzying, thrilling ride into the system... just seeing Pluto there getting bigger and bigger, it gives me goose bumps."*

Perhaps for the last time in a generation, the textbooks on Pluto were rewritten, virtually overnight. In fact, so profound were the discoveries that one team member inadvertently dropped the "F-bomb" during a broadcast. Smack in the middle of Pluto was a heart-shaped field of nitrogen ice, fed by glaciers made of the gas comprising most of Earth's atmosphere. This field, later called "Sputnik Planum" had no craters, therefore was geologically young and active. How could such an active feature exist on a tiny world whose interior heat should have disappeared long ago? Evidence of erosion and even possible gullying were also present within Pluto's granite-hard ice mountains. Charon also shared Pluto's bizarreness. Instead of wall-to-wall craters, as expected by most researchers, a massive valley sliced through the surface. Charon's north pole also sported an ochre coating, mimicking the brown colour of Pluto. As a final parting shot, New Horizons looked back at Pluto and showed a whisper-thin atmosphere tinted blue. It felt strange that, on the edge of space, Pluto should share a similar Earth-like tint as our own world.

To many, New Horizons strongly echoed the radical exploration of the Voyagers a generation before. Accustomed to being reminded of "missing out" on the Moon landings occurring before they were born, at least one Generation X enthusiast commented: *"We might have missed out on Mercury and Apollo, but we sure didn't miss out on Pluto."*

(Above) After 13 years of continuous exploration of Saturn, dawn arrives for the last time on the aging Cassini. Almost spent of fuel, Cassini prepares to make a suicide plunge into the planet it had spent so long exploring.

Although known since the ancients, telescopic observations of the outer Solar System reframed humanity's place in the cosmos. Galileo's discoveries of Moons around Jupiter kicked Earth from the centre of the universe to one of many planets orbiting around the Sun. Also in the outer Solar System, two spacecraft totally rewrote textbooks virtually overnight.

The exploration of the outer Solar System also revealed how limited human imagination is. Even the far-reaching minds of the early space artists tended to add Earth-like features to their space scenes. The scene in the famous painting of Saturn from Titan could easily have been somewhere in Utah, but for the ringed planet hovering in the sky. Instead, the space age discovered worlds with their own bizarre geology and chemistry, which was alien from Earth. Familiar features such as dunes and glaciers were discovered, but powered by hydrocarbon grains or toothpaste-like nitrogen flows.

The outer Solar System may also hold at least two places that could be potential habitats for extra-terrestrial life. Jupiter's moon Europa, with its subsurface ocean, is inspiring the next generation of spacecraft explorers. An underwater rover is being designed to work underneath the Jovian moon's ice and search the dark mysteries below. Saturn's moon Enceladus also harbours an ocean of liquid water. Even this far from the Sun, could this alien water which Cassini discovered also contain life? The basic ingredients for life seem to be present, but confirmation will need to wait for a future mission.

Despite this interest, the chances of humans venturing out this far are remote. The distances are simply too large, requiring massive expenditures of energy and years of time. For the foreseeable future, such exploits are confined to science fiction, and it is left to robots to continue to explore the worlds of ice.

Memories of humanity's first push out of the Solar System live on, however. Five spacecraft carry with them messages for potential extra-terrestrial intelligence regarding life on Earth. Two of these spacecraft, Voyager 1 and 2, carry gold-plated records on which images, music and messages from a distant Earth are inscribed. These memories provide a snapshot of Earthly existence and theoretically will last long after our Sun has swollen to a red giant and consumed the home world in fire. This may perhaps be the greatest legacy of humanity to the universe, where life on one small planet reached out to the stars.

(Above) Pre space-age artwork of the surface of Io showing mountains and craters (David A. Hardy).

The study of the outer planets tested human imagination to its limits. Their enormous distance from Earth made early telescopic observations difficult, and, thus, Uranus, Neptune and Pluto had to wait until the telescope age to be even discovered. Through telescope, even Jupiter and Saturn, known since ancient times, showed mysteries. Banded Jupiter rotated so fast that it bulged at the equator, while Saturn was surrounded by a ring that had no Earthly equivalent. The very idea that worlds could exist as balls of gas was beyond the furthest flight of fancy for renaissance astronomers. Early astronomers' and artists' perceptions of these planets were very Earth-biased. They imagined larger, more extreme versions of our home planet. Enormous distances and temperatures cold enough to freeze gasses were too fantastic for consideration. Part of the issue was that there was no place on Earth that even remotely resembled the conditions of a gas giant. The environments of Earth were simply not extreme enough for a scientist to gain any context on what the outer planets were actually like. Astronomers settled for the next best thing: volcanoes and wide oceans. In the 1900's, sci-fi stories showcased giant sea monsters fighting in the oceans of Saturn. As late as 1950, Chesley Bonestell and Willy Ley's book 'Conquest of Space' visualised a surface of Jupiter with towering ice cliffs and ammonia lava flows.

As with the inner Solar System, gradually improving telescopes and more robust measurements led to more realistic artistic impressions of the outer Solar System, though until the Space Age there was plenty of room for creative license. This was particularly true for the retinue of moons circling the outer planets like moths around a flame. Appearing as not much more than points of light in even the best telescopes, astronomers could only guess at what they were like. In his artwork for the 1970 National Geographic article

164

(Above) The radiation-soaked, volcanic surface of Io based on Voyager data.

Voyage to the Planets, Ludek Pesek portrayed the surfaces of these moons as cold, cratered bodies, reflecting the light of the distant Sun or their parent planet, echoing the consensus at the time. Almost nobody thought the moons to be more than cold, uninteresting worlds. David A. Hardy painted the surface of Europa using the best information at the time: *"All we knew about Europa in those days (1972 and earlier) was that it has a high albedo, so the assumption was that it has a rocky surface overlaid with ices of various types, and that is what I showed. I assumed it would be cratered because of its proximity to the asteroid belt. It was of course a complete shock when Voyager showed it to be as smooth as a billiard ball and covered in cracks!"*

International Association of Astronomical Artists (IAAA) member Marilynn Flynn explained some of the difficulties in trying to accurately represent the moons of the outer Solar System: *"My space art landscape paintings are always based on current research, but sometimes there just isn't enough information to get the whole story, or there are conflicting views among scientists. In that case, I have to try to find a balance between what we actually know and what we think it might be like. So it's always fun, when a spacecraft arrives at a previously unseen planetary surface, to look back and see if the speculative space art paintings I did before the mission were spot-on or a complete miss. One complete miss was a painting I did of a lake of liquid Nitrogen on Neptune's moon Triton. Before Voyager whizzed past in 1989, scientists had hypothesized that its surface might be covered in frozen methane with a liquid nitrogen lake at the pole. Well the methane and nitrogen are there, but they're both frozen!"*

While the Space Age revealed most of the outer Solar System moons to be bizarrely interesting worlds that initially defied explanation, it had the opposite effect on Saturn's moon Titan. Known to have an

(Above) Before Voyager, most astronomers thought Titan would feature some of the most picturesque views in the solar system. Saturn, drawn into a crescent by the shrunken Sun, would float serenely in blue skies.

(Above) The surface of Titan, sometime in the future. Saturn and the sun struggle through eternal smog while an astronaut surveys alien shores of liquid methane (Michael Carroll).

(Above) Pre space-age artwork of Europa. The little Jovian moon was thought to be a hard-frozen world of ice, relatively unchanged through the eons (David A. Hardy).

atmosphere since the early 20th Century, many artists painted a ringed Saturn hanging in a clear blue sky. Chesley Bonestell's famous painting of Saturn from Titan was one of the iconic images of the 20th Century that was said to have: *"launched a thousand careers in space exploration"*.

Following the mind-blowing images of Jupiter's moons from the Voyager probes, astronomers were eagerly awaiting to be awed by stunning images as Voyager 1 made a close fly-by of Titan. Instead, while technically successful, Voyager's images showed Titan to be a fuzzy, orange ball. The eternal smog perfectly obscured the surface of Titan from Voyager's cameras, and dreams of seeing Saturn hanging in the blue skies of Titan quietly vanished.

Space artists remained interested in the enigmatic moon, however, and the dense atmosphere was thought to hold some more surprises. IAAA member and space artist Michael Carroll described the lure of Titan while creating his work Titan Surf: *"Ever since the 1950's, writers and artists have been depicting Saturn's planet-sized moon Titan in scientifically informed ways. Arthur C. Clarke wrote of Titan's 'methane monsoons' in light of the discovery that surface conditions could hold methane at the triple-point where it exists as liquid, ice and vapour. Hubble studies bolstered this idea in the 1990's. Near-infrared images showed surface variations that indicated large areas of cool terrain, perhaps seas and lakes. Voyager spacecraft could not penetrate the tomato-soup haze of the great moon-world but confirmed that conditions could be ripe for methane in liquid form on Titan's surface.*

The arrival of the Cassini spacecraft, carrying the Titan-bound Huygens probe, heralded a new understanding of the fog-shrouded moon. Huygens returned aerial images during its descent, revealing dendritic river valleys and areas that appeared to be flood plains. From the surface, the probe imaged rounded boulders and stones, eroded cobbles similar to those found in terrestrial riverbeds. Cassini itself imaged river valleys, ponds and lakes, finally confirming the methane rains of this strange world.

(Above) Voyager revealed Europa to be amazingly smooth, having a cracked icy crust sitting above a global ocean. NASA considers Europa to be one of the best candidates of harbouring extraterrestrial life.

The painting 'Titan Surf' could not have been done even twenty years ago, for two reasons: Firstly, our knowledge of Titan was not what it is now — the painting is a close approximation of latest data; Secondly, digital painting was in its infancy. This image is a 'tradigital' painting: it uses a base of traditional acrylic paint on board, enhanced digitally with programs like Photoshop and Terragen.

We see Saturn in the sky, through Titan's thick hydrocarbon fog. Would an astronaut really be able to? The subject has been one of debate for decades. Ralph Lorenz of Johns Hopkins University Applied Physics Lab, a Titan expert, suggests that with just the right polarisation of your sunglasses, Saturn's golden globe would come shining through on the 'clearest' of days. We will just have to await the first visitors to find out for sure!"

Along with the outer Solar System moons, tiny Pluto was historically uninspiring to space artists. Artwork of the tiny planet virtually appeared in monochrome and focused more on the incredible distance of the Sun than the dwarf planet itself. In The Conquest of Space, Chesley Bonestell summed up most of the mid-20th Century knowledge of Pluto by writing: *"Pluto, the outermost planet of the solar system, turned out to be small and massive. Its atmosphere must lie frozen on the rocks. From that distance the sun looks like a brilliant distant arc light, without [a] perceptible disk."*

When United States Naval Observatory astronomer James Christy discovered Pluto's moon Charon in 1978, space artists had the opportunity to paint something other than a tiny Sun in Pluto's skies. Even with the company of Charon, most paintings continued to show Pluto as a cheerless place. Ron Miller and Bill Hartmann wrote of one such illustration in The Grand Tour: 'The Traveler's Guide to the Solar System': *"As pale as a moonlit skull, gray Pluto hangs in the sky of frigid Charon... its giant moon and only companion as it ponderously rolls through the comet haunted darkness at the edge of the solar system."*

In the last planetary encounter of the "original" Solar System, NASA's New Horizons probe flew past

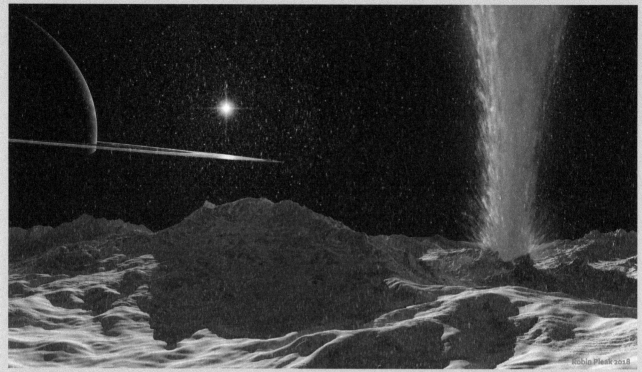

(Above) Erupting geysers create a flurry of snow on Enceladus (Robin Pleak).

the Pluto system and, like Voyager before it, unveiled a surprising new world. Inspired by New Horizon's findings, Marilynn Flynn created Charon from Pluto: *"Charon hovers in the sky over an icy crater on the surface of Pluto. A bright coating of Nitrogen/Methane ice covers an area of older water ice which has been colored red brown by tholins (hydrocarbons).*

On Pluto, frozen water is as hard as rock and forms a sort of bedrock, while Nitrogen, Methane and Carbon Monoxide ices can flow, condense as frost or possibly fall to the surface as snow. Charon is crossed by deep canyons and has a splotch of reddish color at its north pole, possibly hydrocarbons captured from Pluto's atmosphere. Pluto's atmosphere has thin layers of blue haze, and might even have a few clouds. Mini moons Nix (above Charon) and Hydra (below Charon), are icy potato-shaped bodies that tumble wildly in their orbits."

The demotion of Pluto to a dwarf planet was part of the realisation that the Solar System is more complicated than once thought. Pluto is no longer the final outpost from the Sun, and New Horizons is scheduled to reach a Kuiper Belt object, one of many bodies orbiting even further from the Sun than Pluto. As humanity's exploration of the Solar System widens, space artists will continue to play an important role in bringing visions of alien vistas accessible to the public. IAAA artist Ray Cassel summarised the journey so far: *"Years ago it was hard to imagine the golf ball sized methane rain drops on Titan, the blue sunsets on Mars, or the thin blue atmosphere of Pluto. The recent discovery of abundant planets indicates to us that out in the universe there may well be an analog to even the most outlandish ideas we artists dream up."*

(Facing page, top) Charon, the largest of Pluto's moons, floats above an ultra-cold landscape sculpted by ice (Marilynn Flynn).

(Facing page, bottom) The discovery of blue skies and Wright Mons, a possible ice volcano by New Horizons showed Pluto's surface to be far more diverse than expected (Michael Carroll).

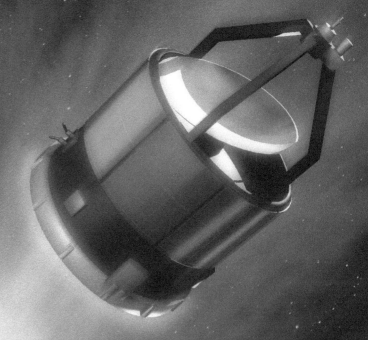

VERMIN OF THE SKY

Early astronomer Johannes Kepler was perplexed at what he saw of the Solar System. After he created his famous laws of planetary motion, Kepler believed there should have been a planet orbiting between Mars and Jupiter. In 1772, Johann Bode came across a formula, later called Bode's Law, that predicted the distances of the planets from the Sun. Bode showed his sequence of numbers aligned with the distances of all the planets from Mercury to Saturn. Like Kepler, Bode was annoyed at the gap between Mars and Jupiter, where his law said there should have been something. He asked: *"Can one believe that the Founder of the Universe left this space empty?"* His question became even more pertinent when William Herschel discovered Uranus at the distance that Bode's Law predicted.

In 1801, Sicilian professor Giuseppe Piazzi found an extra star in the constellation Taurus. Later named Ceres, the 'star' fit almost exactly into Bode's predicted formula, though its tiny size as inferred by Herschel hardly made it a planet worthy of filling the gap. A year later, astronomer Dr Heinrich Olbers discovered a second body — which he named Pallas — orbiting at the same distance as Ceres, and the decades that followed increased the Solar System's celestial inhabitants by hundreds. So numerous came the discoveries in this region, that in 1881 Proctor wrote in The Poetry of Astronomy: *"We have become so accustomed of late to the discovery of planets traveling along the region of space between the paths of Mars and Jupiter, that we are apt to forget how strange the circumstance must have appeared to astronomers at the beginning of the present century, that the old views respecting the solar system were erroneous, and that in addition to the planets travelling singly around the sun the existence of a ring of planets must be admitted."*

Exactly what the "ring of worlds", later called asteroids, precisely looked like remained a mystery. Pondering the question in 1890, R.S. Ball wrote in The Story of the Heavens: *"Of the physical composition of these bodies and of the character of their surfaces we are entirely ignorant. It may be, for anything we can tell, that these planets are globes like our Earth in miniature, diversified by continents and oceans."*

Their small size was beyond the resolving power of Earth-bound telescopes. As the novelty of their discovery wore off, many astronomers grumbled that they got in the way of serious discovery, giving asteroids a new term: *"vermin of the sky."* Very little in the way of artwork or illustration was published, as IAAA space artist Bill Hartmann explained: *"Asteroids in that period were thought of as different from, and unrelated to, comets - which in a way discouraged art related to them, because they would be just static lumps of rock. In terms of "rocks 'n' balls"... they were just rocks, with no ball in the sky to make the picture interesting."*

What little art there was focussed more on the catastrophic effects of meteors raining down on a hapless Earth, wiping out cities or even whole continents. Even the start of the Space Race was not enough to shift the general astronomical apathy of studying asteroids. The real interest they generated in NASA was whether any asteroid belt-crossing spacecraft might suffer an accidental collision with an uncharted space rock. As the Pioneer 10 and 11 spacecraft made the first ever — and safe — passage through the asteroid belt, NASA mission controllers breathed a collective sigh of relief. Asteroids were then promptly forgotten again. After all, why spend millions of dollars and several years to fly all that way, just to visit one, or at most two, crater-ridden lumps of rock?

As serious ideas of rocketry and space travel grew, some artists, helped by science fiction writers, saw another possibility for asteroids; raw material. By this time, asteroids were known to contain an abundance of water ice and other materials that could be used to build and fuel future generations of space ships. Why spend huge amounts of energy building and launching spacecraft from Earth when they could more easily be built in space? Some writers went further, by suggesting using an asteroid itself as a space ship. In 1952, British Interplanetary Society member Dr L.R. Shepherd imagined hollowing out an asteroid and converting it into an ark where generations of people would live out their lives on the way to a distant solar system. Such

(Facing page) The European Space Agency Giotto spacecraft made a once in a lifetime encounter with Halley's Comet in 1986. Returning unprecedented closeup images, Giotto's camera was destroyed by an impact from cometary debris.

(Above) An artist's view of one of the many asteroids in our solar system. Some asteroids, such as the one illustrated here, are accompanied by their own asteroid "moons" (Mark Pestona).

a population would have to be self-sufficient, surviving on what could be grown and carried with them on their interstellar journey. Psychological profiling and group dynamics would have to be carefully planned for this type of journey. Mismanagement of resources, segregation or even battles occurring in such a small space would spell disaster for a potential colony of explorers.

Not all scientists were so dismissive of asteroids. Planetary researchers began demonstrating that craters on the Moon and Mars were made by impacts from asteroid-like bodies. Clearly, asteroids had a role in shaping the Solar System. The discovery of Meteor Crater in Arizona, USA, showed such impacts have also occurred on our planet.

What of the asteroids themselves? Were they the result of an exploded planet, or, as Dr Tom Gehrels suggested, did they actually date from an earlier period? *"The asteroids are probably part of the original nebula... from which we believe the sun and planets condensed some five billion years ago"*, he stated in 1970. If this was the case, then studying asteroids might be the key to unlocking the mystery of the genesis of the Solar System. Eugene Shoemaker, who died tragically in a car crash, devoted much of his life to the study of asteroids. Although health issues stopped him from training as an astronaut, he nonetheless worked hard on learning how asteroids had shaped the Moon. His work with the Lunar Ranger spacecraft helped show conclusively that asteroids, not volcanoes, had formed the Moon's craters.

Thanks to discoveries by Shoemaker and other dedicated scientists, asteroids became more interesting to the space artist, with visualisations appearing in step with scientific discoveries. Bill Hartmann: *"In the '60's the idea of accretion of planets from asteroid-like 'planetesimals' began to take hold offering interesting possibilities... In the '70's there began a classification system according to different spectra and albedo, so that differences that were known could be depicted.*

"In the '80's we began to link asteroids and comets (a continuum based on ice content), with some objects that were originally classified as asteroids were discovered to have cometary eruptions. Also, in '70's

174

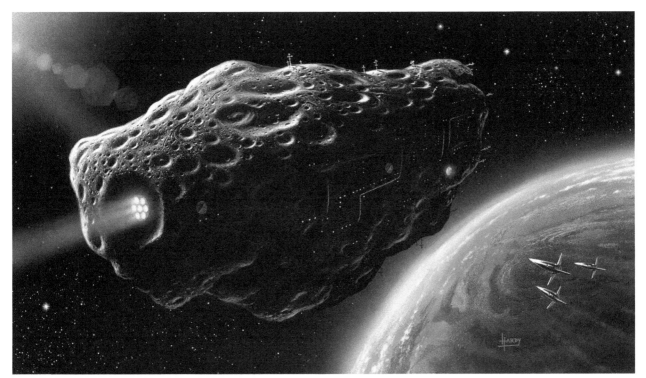

(Above) Sometime, far into the future, asteroids may be hollowed out and converted to interplanetary space ships. Entire generations of people would live their lives within such a craft (David A. Hardy).

and 80's light-curves of asteroids were being observed, showing variations from broadside to end-on view, causing it to be realized that some asteroids are very elongated; some were also found to have satellites.

"I've had fun with some paintings trying to show the view from one end of a slightly banana-shaped asteroid showing the far end looming up in distance over the nearby 'horizon'."

Despite the efforts of Shoemaker, the asteroid groundswell just was not enough to fund a dedicated mission. Humanity's first close-up look at an asteroid had to wait until 29 October 1991, when Galileo flew past 951 Gaspra en route to Jupiter. Already nobbled by a broken high gain antenna, Galileo's cameras transmitted image slivers, that painstakingly built up to full pictures over the ensuing days. Gaspra was a 19-kilometre-long rock, and absolutely full of craters. In fact, Gaspra had more craters than either of the similar sized Martian satellites.

Galileo's second asteroid encounter occurred on 28 August 1993, and was of 243 Ida, showing a fat cigar-shaped mountain of rock, sporting craters on its sides. However, unlike Gaspra, this asteroid held a surprise. Peeking out from behind the larger body was tiny Dactyl, a 1.6-kilometre wide moon, circling the larger parent body. Dactyl's orbit was a precarious one, as the kick of a football from the surface of Gaspra would almost be enough to totally defeat the little world's gravity. A mere wisp of gravity was holding tiny Dactyl in place.

Interest in further asteroid exploration increased in 2003, when a special mission was launched from a new space kid on the block — Japan. Hayabusa was an ambitious mission to approach an asteroid, much like NASA's Near Earth Asteroid Rendezvous - Shoemaker (NEAR Shoemaker) mission launched in 1996. Unlike NEAR Shoemaker, which had simply orbited and then crash-landed on asteroid Eros, Hayabusa would fire a bullet into the surface of an asteroid and return fragments back to Earth. To help the box-shaped spaceship, a small sphere would be dropped onto the asteroid for target practice. Hayabusa also carried a small sub-probe to get up close and personal with the asteroid surface. Although impressive, the sub probe meant to hop along in the minimal asteroid gravity, had replaced a NASA-designed nano-rover. Fitting onto the palm of a hand

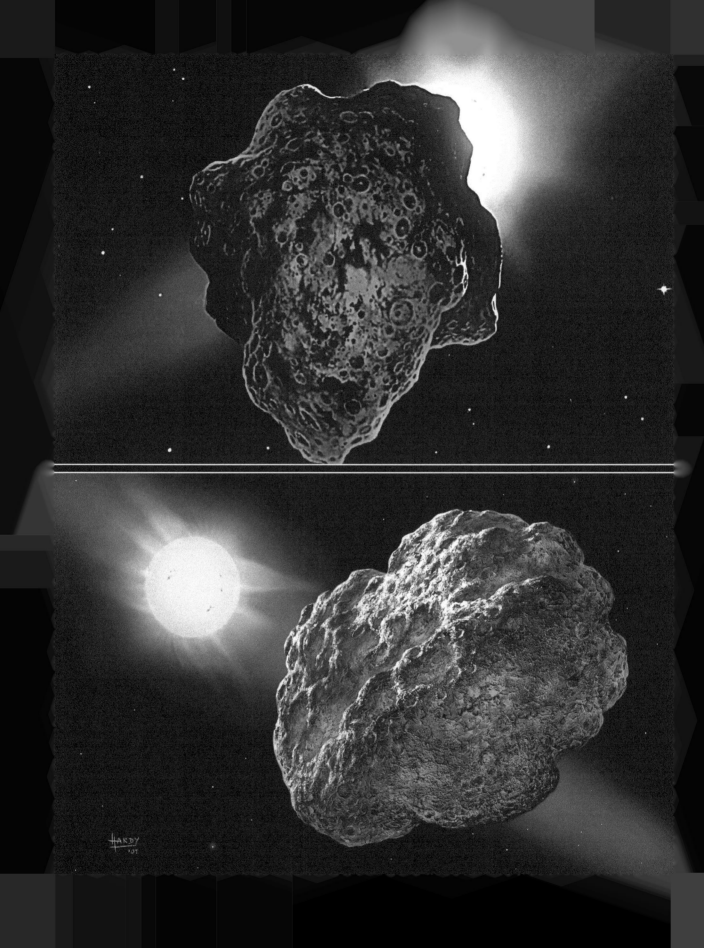

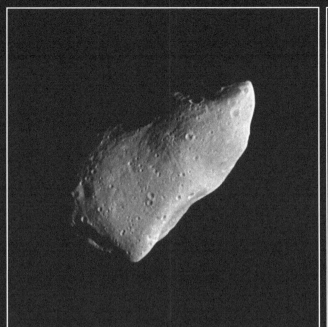

(Facing page) Two views of the sun grazing asteroid Icarus from before (top) and after (bottom) spacecraft provided detailed views of asteroids (David A. Hardy). (Top left) The Galileo spacecraft returned the first ever closeup image of an asteroid (NASA). (Top right) The asteroid Eros as imaged by the Dawn spacecraft (NASA). (Bottom) The tiny moon Dactyl hovers above the horizon of asteroid Ida, held in place by Ida's weak gravity.

(Above) The Japanese Space Agency's Hayabusa probe fired a projectile into an asteroid to gather precious sample material. These were returned to Earth in 2010.

and featuring on a US talk-back show, the tiny rover would have been the smallest wheeled vehicle ever to fly in space.

En route to its destination, the Japanese spacecraft was attacked by the Sun, frying parts of its solar panels and starving its ion thrusters of power. Finally, after a two-year limp, Hayabusa reached its destination, asteroid 25143 Itokawa. Grappling against time delay, Japanese engineers tried to send Hayabusa down to the surface – and lost contact. Rockier than expected, the rough terrain threw Hayabusa into a flat spin. The hapless Minerva sub-probe, prematurely released, flew off into space without ever touching the ground it was designed to explore. A second landing attempt put the spacecraft back into safe mode, and mission scientists had no way of knowing whether Hayabusa's sample capsule contained precious bits of Itokawa, or was empty. A 2008 return to Earth was abandoned following fuel leakage, and Hayabusa was almost given up for dead. Somehow, working on little over half its battery power, Hayabusa and its precious cargo of 1,500 dust-sized asteroid particles, made landfall in the desert of Woomera, Australia, in 2010.

By 2007, other asteroids were finally getting explored. The two largest, Vesta and Ceres, remained unvisited and a mystery. Both worlds appeared to be opposite, with Ceres thought to be a 'wet' or at least an icy world, while 'dry' Vesta was more of a rocky, airless desert. Additionally, as far as asteroids went, both Ceres and Vesta were huge. Vesta accounted for ten percent of the total mass of asteroids, while over a quarter of the asteroid belt's mass was swallowed up by Ceres. Ceres' large size caused an astronomical controversy. Ceres

(Above) Many scientists believe this crater on Vesta holds gullies shaped in part by liquid water. The gullies are located in the upper right quadrant of the crater wall (NASA).

shared more in common with planets such as Mercury or Pluto than with its fellow asteroids. For a short period, the International Astronomical Union (IAU) almost classified the 476-kilometre-diameter world as the Solar System's 10th planet. Astronomy textbooks would have to be rewritten, as school children would have to be taught of a ten-planet Solar System. The situation came to a head with similar size worlds being discovered past Pluto. Should the Solar System consist of 15 planets? Or 20? In 2006, the IAU classed these smaller worlds as 'dwarf planets', and demoted Pluto to this status also. At last, Ceres was accounted for, though bad feelings and even protests over the removal of Pluto as a planet have endured.

The Dawn mission was NASAs effort to study Vesta and Ceres up close. Selected from a series of proposals in 2001, Dawn would use an ion engine to navigate between the two asteroids. For the first time, the ion engine would have to change a spacecraft's velocity as much as the original launch rocket. Typically, the launch from Earth took the lion's share of rocketry effort, with in-space manoeuvres normally requiring modest changes. Dawn was different. It would have to navigate and orbit starship-style around and between two asteroids. Spacecraft navigator Chris Potts explained the difficulties: *"It's quite a different mission. The concerns are a little bit different than other missions because the thrust is so efficient. One of the things you worry about is not having enough time to make any corrections that are needed. Although the thrust is low, over time it builds up quite a velocity change and you're always designing trajectories and changing commands to make sure the ion engine is firing in the right direction. If there's any kind of spacecraft fault or hiccup along the way, you have to scramble, and some future events might have to be moved around."*

(Above) Dawn discovered this massive landslide in Vesta's Marcia Crater. Material from the right has fallen into the crater's interior. Landslides on Earth can cause catastrophic damage to lives and property (NASA).

Ion thrust weren't the only problems to contend with. In 2003, NASA cancelled the mission, briefly reinstated it a year later, and then indefinitely postponed it in 2006. As a last-ditch effort, the company building Dawn offered to sacrifice working for a profit, and NASA finally gave the go-ahead for Dawn. Launching on 27 September 2007, and spending four years travelling on thrust no more powerful than the weight of a single sheet of paper, Dawn arrived at Vesta.

As the spacecraft entered a year-long orbit around the cratered world, Vesta began to throw up surprises. Two massive craters had almost wiped out Vesta's south pole, the largest almost as wide as the asteroid itself. The crater's central peak rose over 25 kilometres high, making it the second tallest mountain in the Solar System. Dawn imaged some of Vesta's younger craters, and a discovery in them was almost the stuff of science fiction. Some of the craters hosted gullies. When found on Earth, these types of gullies are almost always formed by water, and until Dawn, only seemed to occur on one other body where liquid water had previously acted: Mars. Quite modest by Martian, or even Earthly standards, these 30-metre-wide channels snaked down the crater wall to end in a fan-shaped deposit. Postgraduate researcher Jennifer Scully was surprised at the implications of gullies on Vesta: *"Nobody expected to find evidence of water on Vesta. The surface is very cold and there is no atmosphere, so any water on the surface evaporates. Vesta is proving to be a very interesting and complex planetary body."*

Scully continued: *"We're not suggesting that there was a river-like flow of water. We're suggesting a process similar to debris flows, where a small amount of water mobilizes the sandy and rocky particles into a flow"*. The idea of even tiny amounts of liquid water on a body much smaller than our airless Moon, and further from the Sun, was revolutionary. Dawn's Principal Investigator Christopher Russell said: *"These results, and many others from the Dawn mission, show that Vesta is home to many processes that were previously thought to be exclusive to planets."*

Similarly, the dwarf planet Ceres was more than it seemed. Earlier space art of the largest object in the asteroid belt showed Ceres to be a grey ball, smattered with craters. Pre-Voyager thinking had dismissed

(Above) A false colour view of one of the mysterious bright spots in one of Cere's craters. These bright spots may be salt deposits, or even remnants of past icy volcanism (NASA).

Ceres, and other smaller worlds such as Jupiter's satellites, as uninteresting, dead places. Dawn revealed smatterings of ultra-bright deposits within Ceres' craters. Although conspiracy theorists proclaimed these as evidence for alien cities, NASA scientists were just as excited when Dawn's data suggested that there were salt deposits. Salt meant that Ceres once had lots of water earlier in its history. Remnants of this early water may have even driven icy volcanism in its recent history. Could Ceres still be active? Could life have been possible on Ceres? As is usual in science, Dawn answered old questions and raised new ones.

Marc Rayman summed up the Dawn mission so far: *"Scientists are still in the initial stages of digesting the sumptuous data Dawn has served from Vesta and Ceres. There is already some evidence, however, that dwarf planet Ceres did not form where it is now. It may have formed even farther from the Sun than Jupiter currently resides. If so, when scientists piece together the story of how it was relocated, it surely will provide more information about how the planets jockeyed for position in the early solar system. Protoplanet Vesta, bearing a closer resemblance to the terrestrial planets than to the much smaller chunks of rock we call asteroids, may be showing us an example of the major building blocks of planets, including Earth."*

As with NASA, space artists are still digesting Dawn's unexpected discoveries of the complexity of Vesta, Ceres and other smaller bodies. Dawn and other asteroid exploratory missions are also reviving the science fiction idea of using asteroids for resources in space. While construction of such asteroid ark ships probably resides far in the future, NASA has been developing asteroid mining for more near-term goals. As Dante Lauretta, Principal Investigator of NASA's future sample return mission, explained: *"Water is a critical life-support item for a spacefaring civilization, and it takes a lot of energy to launch it into space. With launch costs currently thousands of dollars per pound, you want to use water already available in space to reduce mission costs. The other thing you can do with water is break it apart into its constituent hydrogen and oxygen, and that becomes rocket fuel, so you could have fuel depots out there where you're mining these asteroids. The other thing C-type asteroids have is organic material – they have a lot of organic carbon, phosphorous and other key elements for fertilizer to grow your food."*

NASA is even fast-tracking a pioneering mission to look at a potentially mine-worthy asteroid, 16 Psyche. Studies from Earth showed the 210-kilometre-diameter asteroid to consist almost entirely of iron and nickel. In fact, the amount of metal in the asteroid, if brought to Earth, would totally crash the global economy many times over. Fortunately, the Psyche mission, scheduled for launch in 2022, has instructions to only investigate the controversial asteroid, not return anything home. Results from the Psyche mission may fuel interest in asteroid mining. Private companies are looking at sending prospecting spacecraft by the end of the decade, while the Obama Administration passed a law allowing US citizens to own materials from asteroids they mined. Whether the current generation of spacecraft is sufficient to mine asteroids, or even if humanity agrees whether it is ethical to mine them, remain questions for the future.

Unlike asteroids,
comets haven't suffered the obscurity crisis of their rocky cousins and appeared many times in historic artwork. Sporting feathery tails behind them, comets were once thought to be omens of doom. The fate of kingdoms, battles and royal lineages hung in the balance if ever a comet appeared in the sky. More recently, most scientists believed that a cometary impact with the Earth helped wipe out the dinosaurs. Comets inspired artists over the centuries, with Italian artist Giotto di Bondone panting a comet in his 1304 painting Adoration of the Magi. Various space artists have also visualised what it would be like for a person to stand on the surface of a comet as the ground literally boiled away around them.

When Isaac Newton gifted the world with his theories of gravitation, 17th Century astronomer Edmond Halley wondered whether comets might move in elliptical orbits around the Sun. Halley's painstaking research into comet records showed that he was on the right track. In 1682, a bright comet appeared in the skies above Europe, and Halley used his newfound knowledge to confirm it was following an elliptical orbit around the Sun. Working backwards, he discovered many comet sightings from history, including artwork on the famous Bayeux Tapestry, were in fact the same celestial object. Halley died before the comet named after him again visited the Solar System. In 1910, the first ever photographs of Halley's Comet burst onto Edwardian society, and by the time of its next encounter in 1986, the Space Age was well underway.

In the years leading up to the 1986 encounter, a host of nations, including the USA, Russia and countries in Europe, began feverishly designing missions to meet the once-in-a-lifetime visitor. Giotto, named

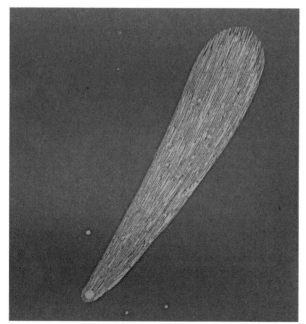

(Above) A 19th century illustration of a bright comet (Public domain).

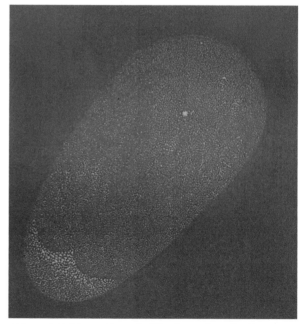

(Above) William Herschell's view of Halley's Comet (Public domain).

(Above) In 1882 a majestically bright comet lit up the skies above Victorian England (Public domain).

(Above) The misshapen nucleus of Halley's Comet is seen for the first time by the Giotto probe. Giotto's success would lead to the Rosetta mission decades later (ESA).

after the Italian artist who illustrated the comet for the Star of Bethlehem in 1304, was ESA's first ever deep space mission. The mission almost died on the drawing board, when the USA suddenly pulled out of the Halley race, along with its share of funding. Russia and Japan came to the rescue, saving the mission.

As Halley's Comet brightened the skies over Earth, an armada of Russian spacecraft, followed by Giotto, blazed a trail to their goal. On 13 March 1986, Giotto flew perhaps the most risky profile of any primary mission before it. Creeping up on the comet's nucleus, the little spacecraft was blasted by tail dust travelling over 60 times faster than a rifle bullet. Cameras on board Giotto returned humanity's first ever look inside a comet, showing a 10 by 15 km coal-black nucleus spewing out gas from its boiling surface. Suddenly, just before closest approach, a particle not much bigger than a grain of sand sent the hapless spacecraft reeling. The camera, just recently returning those historic nucleus images, was destroyed instantly, while for the next 30 minutes, Giotto fought for its life. The now blind spacecraft eventually regained contact with Earth, and even survived to encounter another comet in 1992. In the meantime, ESA had proven to the world just what it could do in space. Former deputy project scientist Gerhard Schwehm stated: *"It was a once-in-a-lifetime event and it had a big impact on the general public... Giotto ignited the planetary science community in Europe – we had demonstrated that we could successfully lead demanding missions – and people started thinking about what else we could do."*

From an earthly perspective, Comet Halley's 1986 visit was almost a bust, particularly due to the hype space agencies had generated about it the year before. Inspired by historic photos of Halley's brilliant form in 1910, thousands of amateur astronomers awoke well before sunrise to view what they hoped was the wonder of a lifetime. Instead, a hazy ball, barely visible to the naked eye and not much better in binoculars, presented itself in the pre-dawn sky. Disparaging comments from *"where's the rest of it?"* to *"Better luck next time,"* followed Halley's Comet as it left the inner Solar System for deep space.

Despite Comet Halley's lacklustre performance through the telescope, Giotto's incredible discoveries inspired NASA to finally design missions to explore comets. In 1999, NASA launched Stardust, a 390-kilogram probe that would not only chase down a comet as Giotto had done, but bring samples of it back to Earth. *"Stardust is out to answer very very cool questions about the formation of the solar system and the formation of Earth,"* said Project Manager Tom Duxbury of the mission. Following a billion-kilometre space chase that took in a flyby of Earth and an encounter with asteroid 5535 Annefrank in 2002, Stardust bore down on its quarry, Comet Wild 2.

"The comet encounter was tremendously exciting," said Principal Investigator Don Brownlee. *"The biggest problem was that we were a sample collection mission and we wanted to collect a large number of very small particles. But we did not want to get hit by a big enough particle to destroy the spacecraft."* To make matters worse, Wild 2's coma would make precise navigation impossible, and there was a chance that Stardust could run into the very nucleus it was trying to study. *"We didn't even know where [the comet] was precisely until we used our on-board camera to measure the position,"* said Brownlee.

Remembering the near-disaster that struck Giotto, nervous NASA engineers tweaked Stardust's planned trajectory, moving it to a safer distance away from Wild 2 in case of mishap. Then, in January 2004, Stardust flew to within 240 kilometres of Comet Wild 2's nucleus, half the distance of the Giotto encounter. *"Things couldn't have worked better in a fairy tale,"* said Duxbury of the encounter. *"No spacecraft has flown so close to a comet and survived perfectly,"* said an impressed Brownlee. *"The flyby was a fabulous success and we were stunned by the data we collected on board."*

Two years later, the Stardust sample return capsule, containing some of the most valuable material in the Solar System, touched down in the country that launched it. Navigator Chris Potts was there to witness

(Above) In 2005 NASA deliberately sent a 300 kg copper missile to crash into a comet. Deep Impact has just released its subprobe and quickly changed course to stop itself following its missile to impact.

185

(Above) The moment of impact by Deep Impact's sub-probe into Comet Temple 1 (NASA).

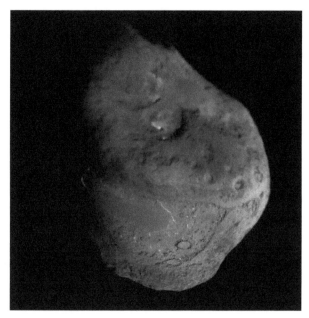

(Above) An extra crater (middle bottom) appeared on the nucleus of Comet Temple 1 (NASA).

the historic landing. *"With the Stardust sample return, to watch the capsule land right where it was supposed to in Utah was very rewarding," he said. "And to see the scientists get their hands on that data and start to perform their investigations, you sense how thrilled and excited they are to finally get to work on their lifelong ambition."*

Just as Stardust's capsule was taking its long journey home, another NASA mission was preparing for the most ambitious comet mission yet. In January 2005, Deep Impact launched from Florida on one of the most ambitious and controversial space missions in history. For the first time, a spacecraft would be deliberately crashed into a comet to study the underlying surface. Deep Impact was planned during a time of increased awareness of how much humanity had been impacting the environment around Earth. From land clearing to producing excessive greenhouse gasses, an environmental movement was rising, and to them Deep Impact's purpose was cosmic vandalism. Would this mission be the start of destructive exploitation of space as humanity was no longer limited to simply observing the cosmos? NASA was more positive about its mission, focussing on the scientific return from such a venture. *"Understanding conditions that lead to the formation of planets is a goal of NASA's mission of exploration,"* said NASA executive Andy Dantzler. *"Deep Impact is a bold, innovative and exciting mission which will attempt something never done before to try to uncover clues about our own origins."*

"From central Florida to the surface of a comet in six months is almost instant gratification from a deep space mission viewpoint," said Deep Impact Project Manager Rick Grammier. *"It is going to be an exciting mission, and we can all witness its culmination together as Deep Impact provides the planet with its first human-made celestial fireworks on our nation's birthday, July 4th."*

On 4 July 2005, a 370-kilogram copper missile separated from its mother craft and guided itself on a collision course with Comet Tempel 1. *"In the world of science, this is the astronomical equivalent of a 767 airliner running into a mosquito,"* said Deep Impact mission scientist Don Yeomans. *"It simply will not appreciably modify the comet's orbital path. Comet Tempel 1 poses no threat to Earth now or in the foreseeable future."*

Automated guidance systems on the impactor steered its way to the surface of Tempel 1 better than any human pilot ever could. Slamming into it at over 37,000 kilometres per hour, the impactor gouged a

(Above) In 2007 the Rosetta spacecraft passed within 1000 km of Mars in order to pick up speed to reach its target comet. Parts of Mars and the spacecraft's solar panel are visible (ESA).

crater the size of a football field, while the mother craft crammed as much data gathering as possible into the 13 minutes it had before being awash with collision debris. Watching the imagery of the impact plume, JPL Director Charles Elachi stated, *"The success exceeded our expectations."*

As it turned out, although the public was bedazzled by images of the blinding flash of the equivalent of a five-kiloton nuclear warhead, scientific data of the actual crater left behind by the impactor were terrible. NASA had to re-task Stardust, fresh from returning a comet sample to Earth, to investigate Deep Impact's comet.

Finally receiving reasonable quality images from Stardust, NASA was surprised at what they found. In a press release, University of Maryland representatives said, *"Measurements at the impact site suggest that the nucleus of Tempel 1 is at least 75 percent empty space, or about as fluffy as freshly fallen snow."* It was,

(Facing page) Rosetta reached comet 67P/Chuyumov-Gerasimenko, surrounded by tendrils of escaping gasses, in 2014. (Top left) Rosetta on approach to the dumbbell-shaped comet nucleus (ESA). (Top right) Jets of gas escape from the comet's nucleus (ESA). (Bottom left) A closer view of 67P's nucleus (ESA). (Bottom right) Icy cliffs tower over the surface of the comet where snow, and recently, dunes, have been discovered (ESA).

according to a senior NASA scientist, as if the mosquito had actually gone through the windshield of the 767 jet instead of just going 'splat'. Comets were revealing themselves to be complex, and somewhat mysterious beasts.

(Above) A view of Philae's landing site as it descended towards its historic landing (ESA).

On 2 March 2004, an ESA mission was launched that was perhaps the biggest gamble since Giotto's encounter with Halley's Comet. The 1.25-billion-dollar Rosetta mission hoped to achieve what no other spacecraft had done before it: land on the surface of a comet. Only envisioned by the occasional space artist, cometary surfaces promised to be a dangerous mixture of boiling gasses and shifting ground. A lander would not only have to contend with spurting jets of debris, such as those that had damaged Giotto, but also virtually non-existent gravity. Such a spacecraft, later named Philae, assuming it survived the approach to a comet, would have to find a way to stick itself on the ground in very low gravity before it floated off into space. Such a fate had already befallen one of the Hayabusa sub-probes, where a small navigation error resulted in the probe drifting away forever.

The other challenge was the mission itself. Rosetta would have to travel in deep space for a decade before even getting a chance to reach its cometary quarry. To save money, Rosetta would be switched off and placed in hibernation mode, to be hopefully awakened before the 2014 comet encounter. Former deputy Project Scientist Gerhard Schwehm was worried: *"You don't want in your wildest dreams to be sitting there in 2014 and the little beast doesn't switch on."* Additionally, a billion Euro gamble had to be performed at Mars as Rosetta would be forced to rely on batteries as it passed through the planet's shadow.

Two years after launch, less than a fifth of the way into the mission, Rosetta's critical fuel supply began to leak. Hasty workarounds ensured that the ailing spacecraft would have enough fuel to finish its mission. Other heart stopping problems occurred as not one, but two of the four critical reaction wheels began to overheat just before Rosetta had to shut down and hibernate. Using a similar process used for the ailing Voyager spacecraft decades before, the Rosetta team frantically created in-flight workarounds to save their spaceship. Then, one by one, all flight systems were shut down as Rosetta entered hibernation and became the furthest human-made object ever to operate on solar power alone. *"We've planned for hibernation for some time, and it's a complex phase of the mission,"* said Spacecraft Operations Manager Andrea Accomazzo. *"Still, for the flight control team, it's an emotional moment. We're essentially turning the spacecraft off. We're already looking forward to January 2014 when it wakes up and we get our spacecraft back."*

On 6 August 2014, Rosetta, which was gingerly woken up the previous January, began to close with what it came all this way for. Comet 67P/Churyumov-Gerasimenko was a dumbbell-shaped hunk of rock and ice, slowly rotating in space. To the scientists, witnessing a never before seen cometary surface was fantastic; to the engineers, Rosetta's cometary images were a disaster. Somehow, they had to figure out how to send their once in a lifetime Philae lander safely down to the surface of that impossible landscape. Comet 67P/Churyumov-Gerasimenko's irregular shape made its miniscule gravity unpredictable, and the hapless Philae could easily be swung off course and miss its target.

As Rosetta spent the next two months entering orbit around the nucleus, engineers pored over the returned images to pick the best site for their lander. *"As we have seen from recent close-up images, the comet is a beautiful but dramatic world – it is scientifically exciting, but its shape makes it operationally challenging,"* said lander manager Stephan Ulamec, understating the problem. Using a similar process used for choosing Martian landing sites, Philae's landing team found five candidate sites that were whittled down to a Site J on the head of the comet. Just four kilometres long at its widest point, Site J was a compromise between getting Philae down safely and the lander actually seeing anything interesting when it got there. Confident that they had done all they could, the Rosetta team issued the command for the 100-kilogram Philae lander to be sent on its historic encounter. Gliding through space, Philae's outstretched lander legs touched the cometary nucleus, and bounced. Harpoons failed to anchor to the ground and the hapless Philae bounced away, its own momentum easily overcoming the ultra-weak gravity. When the lander finally came to rest, it was in the worst possible place, practically lodged in a permanently shaded crevasse. As mission controllers started cheering with confirmation of the successful landing, they knew there would be no opportunity to recharge the landers' dwindling power supply.

Racing against time, Philae returned images of a bizarre alien surface. Space artists such as Ron Miller and Bill Hartmann had painted what they thought were strange cometary vistas; Philae showed that comets were stranger still. *"What really blows my mind is to have this combination of complimentary results,"* said project scientist Nicolas Altobelli. The surface of Comet 67P/Churyumov-Gerasimenko was much more diverse than imagined, with fractured, bouldered terrain interspersed with drifts of soft material. Organics abounded, exciting scientists to the possibility that comets could have seeded the solar system with precursors to life.

Two years after Philae fell silent from lack of power, the Rosetta mother craft was itself dying. Comet 67P/Churyumov-Gerasimenko was drifting further from the Sun, starving Rosetta of power. In one last major manoeuvre, mission controllers sent Rosetta on a gentle but final landing on the comet it had studied for so long. Following the path of its lander, Rosetta continued returning images right to the end. Finally, operations manager Sylvain Lodiot confirmed the end: *"All stations and briefing room, we've just had loss of signal at the expected time... This is the end of the Rosetta mission. Thank you and goodbye."*

Somewhere, far out in space, a largely inactive Comet 67P/Churyumov-Gerasimenko continues on its journey to the dark corners of the outer Solar System. The increasingly feeble sunlight occasionally glints off shiny, metallic surfaces of the long dead Rosetta and Philae probes. These artifacts, perhaps never again to be seen by human eyes, remain evidence of a time when ESA fulfilled its promise of the Halley's Comet days and reached out to touch a comet.

(Above) Although operating for just two days, the Philae lander returned unprecedented images of its landing site on comet 67P/ Chuyumov-Gerasimenko (ESA).

BEYOND

The study of our planets is ongoing. For over a generation, the USA has dominated lunar and planetary exploration. In recent years this has changed. Other nations have become serious contenders in space exploration, with not only Europe, but China and India launching substantial missions into deep space. In fact the only Moon landing to have occurred in the last 40 years was not a human, was not from the US, but a robot from China. Although the US has since renewed its interest in returning people to the Moon, nothing is certain. The last few decades have certainly seen profound advances of our understanding of our neighbouring worlds. Reminiscing of life before the space age, one person who lived through the experience recalled: *"At school back in the '50's we knew the Earth went around the Sun, and the names of the planets but that was pretty much it. Most people were struggling as it was to earn enough money to survive, and there wasn't enough time to learn much more. In fact, many of the farm labourers at the time still thought the world was flat and worried about falling off the sides."*

Dawn mission director Marc Rayman also reflected on the profound change the space age has made on our understanding of the solar system: *"Before the space age, the outer planets were little more than fuzzy blobs, with few clear features plus a host of indistinct moons. Their study was the domain of astronomers. Indeed, planetary science was a small field. Thanks to the spectacular advances in the space age, we now have richly detailed, intimate portraits of the planets and their moons. Now planetary science is a broad field and the outer planets and moons are studied by not only astronomers but also scientists specializing in atmospheric chemistry and dynamics, magnetic fields, solar system dynamics, planetary formation and evolution, geology, and many other fields."*

Throughout this journey of discovery artists have helped exercise vision to anticipate and plan for future missions to the planets. Although gigabytes of hard data from dozens of space missions fill archives every day, space artists continue to imagine the planets. Many of these works have been used to generate publicity needed to sell proposals to the public. Still other artists are actually using spacecraft data and reprocessing it to generate never before seen images. There is simply too much data to be processed professionally, and a growing army of artists, citizen scientists and enthusiasts are wading through the archives and pulling out imagery that often leads to new discoveries. Reprocessing of Voyager imagery showed that the water-rich geyser spurts on Enceladus were actually seen by the Voyager spacecraft, decades before Cassini. Reprocessing of decades old data on Saturn's moons Tethys and Dione revealed new and unexpected geologic features.

The question of 'are we alone?' has fuelled much of astronomy and efforts on the space race. Early astronomers discovered that the sun was a star, and the heavens were filled with other stars too numerous to count. Surely at least some of these stars, contained in myriads of galaxies, would also have solar systems, some like our own? Telescopic observations gave mysterious clues, with some stars, such as Bernard's Star, appearing to wobble in their motion. It was as if unseen companions were pulling them ever so slightly out of true.

Painting worlds that may not have ever existed, while still attempting to be scientifically accurate was a challenge to many space artists, such as Chesley Bonestell. Then, as now, artists began with using what was known about the star system their hypothetical planet was to orbit. This would give them star colour, apparent size, and help estimate how far away a planet would need to be to support life, if that was the artist's intention. Working out the composition or surface conditions of such a planet could only be imagined. Many artists used Earth-based landscapes as starting points for 'filling in the blanks', adapting a terrestrial mountain chain or volcano to an alien setting.

In 1990 the first hard evidence for the existence of extrasolar planets came from an unlikely source.

(Facing page) A ringed gas giant, not unlike Saturn, orbits an alien sun. In the foreground one of its moons hosts rugged mountains, oceans of water and possibly life. Thousands of exoplanets have been discovered since 1990.

(Above) Aurorae flash around the pole of an exoplanet orbiting the active star Tauri V1. Rings of dust surround the planet (Dr Mark A. Garlick).

(Facing page) A view from the surface of Trappist-T1, one of seven potentially habitable worlds discovered around one star (Dr Mark A. Garlick).

While studying pulsars astronomer Alex Wolszczan saw one whose orbit wasn't quite right. Known for their regular pulsating of energy that gave pulsars their name, this particular one had the equivalent of an irregular heartbeat. Wolszczan found the best explanation to be if the pulsar was orbited by two planets, throwing it slightly out of sync.

Wolszczan's discovery, and ones after it showed the Sun's family of planets is not unique in the universe. Humanity discovered first one, then two, and now thousands of exoplanets. In a similar way to the bizarre imaginings of the renaissance telescope observers, the realm of exoplanets is where imagination comes into its own. Exoplanets cannot yet be imaged directly, and most are discovered by looking at wobbles of their parent star, or the ever so slight dimming of starlight as a planet passes in front of its sun. NASA's Kepler Telescope, launched on 7 March 2009 revolutionised the search for exoplanets using the star dimming technique. Able to scan 165,000 stars at once, data from the telescope suggested that every star in our Milky Way galaxy has at least one planet. If confirmed, even pessimistic views estimate a billion Earth-like planets capable of supporting life exist in our galaxy alone.

In 2010, a citizen science project called Planet Hunters took advantage of Kepler's freely-available mission data to open the doors of planetary discovery to anyone. Caltech staff scientist Jessie Christensen explained: *"People anywhere can log on and learn what real signals from exoplanets look like, and then look through actual data collected from the Kepler telescope to vote on whether or not to classify a given signal as a transit, or just noise,"* says Christiansen. *"We have each potential transit signal looked at by a minimum of 10 people, and each needs a minimum of 90 percent of 'yes' votes to be considered for further characterization."* So far Planet Hunters discovered over 2015 planets ranging from Earth-sized to Jupiter sized, as well as the discovery of a five-planet solar system.

The Kepler mission not able to directly photograph exoplanets, and the more advanced, yet to be

Aurorae dance in the sky of an exoplanet discovered orbiting Proxima B. Proxima Centauri is the closest star to the Earth, just over four light-years away. Our sun would be a bright star shining in this planet's night sky (Dr Mark A. Garlick).

The quality of amateur astronomer images exceeds what the best telescopes were able to achieve just before the space age. Our nearby Moon (top left, top middle) remains a popular target for amateurs. Ultraviolet filters show detail in the clouds of Venus (top right). (Middle row) These amateur Mars images reveal clouds hovering over the polar caps, the giant Tharsis volcanos and Vallis Marineris. (Bottom left to bottom right) Amateur astronomers play a key role in monitoring changes in the weather of outer solar system gas giants, such as Jupiter, Saturn and Uranus. (All images Anthony Wesley). (Overleaf) A partially eclipsed Moon hangs over Luna Park in Sydney, Australia (Trent McDougall).

launched James Webb Telescope will only be able to directly image light from Jupiter sized worlds. Apart from broad scale measurements of the planet's atmosphere, and detecting the possible presence of water, detailed views of these mysterious worlds lies far into the future. What little we do know of them is utterly mind-bending. Many of the planets orbit so close to their star that a whole year lasts a matter of days, and wind blowing in their superheated atmospheres roar faster than the latest fighter jets. Others are shedding outer layers of gas, while also boiling carbon and oxygen, elements critical to life. Still others exist within the 'goldilocks zone', regions surrounding the parent star where temperatures are just right for life as we know it. These later planets are the most exciting and it may only be a matter of years before the James Webb Telescope finds indications of a biosphere.

Sometimes artists shy away from adding too much detail, for fear of inventing something that may not be there. The viewer's imagination is left to fill in the blanks. Other artists dived in, to imagine fantastic alien vistas, sometimes complete with alien life. Even with the impending launch of the James Webb Telescope, NASA's most sophisticated (and expensive) telescope designed to actively image the ultra-dim planets, the space artist's role is undiminished. IAAA president Jon Ramer explained: *"Human beings are visual creatures. We learn from looking at the world around us. And we teach by creating images of what we have seen. The painted works of artists reach across the ages, no matter the culture or language of painter or viewer, art speaks to the very core of what it means to be human. As humanity's technological prowess increases, so does our ability to see and imagine new worlds, and our desire to go to those far off places. The power of human imagination has only soared off our planet in the last couple of hundred years. Stories and art depicting other places in our solar system have grown from rudimentary drawings like that of Paul Dominique Philippoteaux's depiction of Saturn's rings for Jules Verne's 1877 story "Off on a Comet" to images of the rings of Saturn based upon actual pictures from probes.*

Today, we dream of far away galaxies and strive to make those dreams come true. We build amazing probes, telescopes, and antenna that are revealing the true nature of the amazing universe around us. And as each new secret is discovered, artists are there to show us more than what cameras can see, helping us to better understand what strings of ones and zeros mean, imagining that far away universe and creating the one thing that humans need to understand, visual images. Art inspires the human spirit and space art inspires us to reach farther than we've ever reached before. Someday we will travel the stars in person, until then we do it through the mind of an astronomical artist."

Although the space age has filled society with close-up imagery of alien worlds, the romance of scanning the skies with a backyard telescope has never gone away. The nearby Moon remains an attractive target, with modest telescopes revealing the Lunar mare and craters that Galileo saw centuries ago. Further afield, amateur astronomy affords regular observations of the planets, particularly in the outer solar system, where professional telescope time is limited. Amateur astronomers also regularly contribute to discoveries that their more complete telescopic gazes can afford.

In 1994, comet Shoemaker-Levy 9 broke up and carpet bombed Jupiter, being proclaimed as a once in a lifetime planetary event. Fifteen years later, Australian amateur Anthony Wesley saw that Jupiter had been hit again. A dark smudge that would have covered the Pacific Ocean swung into view near Jupiter's south pole. Following this and subsequent Jupiter impacts, NASA has set up a veritable hotline where amateurs can provide recordings and data to help advance planetary science.

Amateur astronomy is increasingly being democratised by ever cheapening digital equipment. Digital cameras, wavelength filters and spectrometers allow amateur astronomers to gather scientifically useful imagery and data. Continuous observations on the ever changing cloud belts of Jupiter, or the distant aqua blue planet Neptune help fill in gaps left by the sporadic observations of professional instruments. Amateurs have also made their mark on discoveries of those fleeting visitors to the inner Solar system: comets. Almost a

(Facing page) An asteroid passes over the pole of Uranus on its lonely journey through the cold reaches of the outer solar system. Although visited in 1986 by Voyager 2, there are no plans for a follow-up mission this far from the sun.

(Above) A boulder stands sentinel over the desolate moonscape in this Apollo image. The blue marble Earth hangs in the jet-black sky (NASA).

decade after the excitement and anticlimax of the arrival of Halley's Comet, two amateur astronomers discovered a hazy blob in a star cluster that shouldn't have been there. Alan Hale and Thomas Bopp realized they were seeing a comet, streaking towards an encounter with the inner solar system. In 1997, comet Hale-Bopp outshone the more famous Halley's Comet many times over, and appeared large enough to cut through light pollution of capital cities.

The potential for headline-making discoveries notwithstanding, amateur astronomy also enables everyday people to connect to space. Few forget the first time they saw the rings of Saturn through a telescope. Amateur astronomer Julie Trent described the experience : *"Astronomy has always held an interest for me, but I could never have guessed just how greatly it would affect my life. Growing up in a small town, there was little access to telescopes or people who were knowledgeable about such things. So some time after I'd moved to a large city, I jumped at a chance to attend the local astronomy club's public open night. I saw Jupiter in all its glory that night, with clear orange bands and four little moons. It was beautiful and I wanted more, so I suggested my boyfriend and I go halves in a 5" reflector telescope and join the astronomy club as a Valentine's gift to each other.*

Amateur astronomy can get as complicated as you wish, or remain as simple as looking up. One of my favourite things to do is watch for meteors – no equipment required, no cost, no time constraints. Another favourite is public outreach. Many times, my husband and I have participated in astronomy open nights, sidewalk astronomy events, or just set up the scope in a park and invited passers-by to have a look. It is almost guaranteed to put a smile on people's faces, especially ours."

As Apollo 17's Lunar Module prepared to leave the Moon for the last time on 14 December 1972

NASA engineer Ed Fendell was nervous. His job was to control the Lunar Rover television camera. Parked at a safe distance, his plan was to pan and tilt the camera to capture the ascent stage launch. The few second time lag between the Earth and the Moon complicated things, making controlling the camera not quite 'live'. The troublesome process had been tried twice before with Apollo 15 and 16, with limited success. Fendell had this one last chance to get it right.

Following a final countdown, Gene Cernan fired the Ascent Stage engine and the rover camera captured the moment of ignition, and gracefully panned to follow the two Moon explorers rise off the Moon to re-join their comrade orbiting in the Command Module high above. As the astronauts finally left the camera's field of view, Fendell, satisfied of a job well done, panned back down to film the now deserted Taurus Littrow site. Dominating these last images of the Moon's surface from Apollo was the silver and gold descent stage, resting where it was when Apollo 17 first landed on the grey Lunar dust. This representation of humanity's last sojourn to another world remained with five others just like it around the Moon's equator.

The Lunar Rover camera finally succumbed to failing batteries and a little while later Apollo 17 splashed down in the Pacific; global attention had already shifted elsewhere. Although many promises of a return to the Moon were made by various heads of states in the decades since Apollo, the Lunar surface remained unvisited and still.

Apollo, as with India's Mars Orbiter Mission and other space exploration projects, inspired generations of future scientists, engineers and philosophers. Discoveries from exploring space also directly benefited work in monitoring climate change on our own planet. Remote sensing satellites that monitor carbon dioxide and temperature changes around Earth had their legacy in the space race.

While robotic space exploration continues, space artists also press ahead with their work. After all, new discoveries do not mean much to the public if they can't see it as if they are there. Even as torrents of data arrive from deep space, the artist's ability to turn scientific findings into something attractive to a general audience remains relevant today.

Artwork can also be a snapshot in time. The iconic art of volcanos on the Moon or jungles on Venus are a time capsule of humanity's dreaming of other worlds, as well as a record of imagination based on the best information available at the time. Over time, the sketches and paintings allowed people to begin treating our neighbouring worlds as real places, as much a part of reality as places on Earth.

At the start of the space race NASA saw a future of space art and then Administrator James Webb began commissioning fine arts work. In 1963 he announced to the press: *"An artistic record of this nation's program of space exploration will have great value for future generations and may make a significant contribution to the history of American Art."*

As we move to the next 50 years after Apollo, the difficulties in sending humans to other worlds besides Earth are more political than technological. Most of the Moon walkers have since passed into history, though new generations of explorers continue to explore the worlds around us through robotic proxies. Elon Musk and other private entrepreneurs also promise to change the course of space exploration. Also, more than ever, citizen scientists and enthusiasts are seriously contributing to planetary science. All the while, space art continues to imagine a reality that is much grander in scale than anything most people will ever experience.

(Above) Artwork such as 'The Wave', the first watercolour ever painted in space, may inspire a whole new genre of literal 'space' art (Nicole Stott).

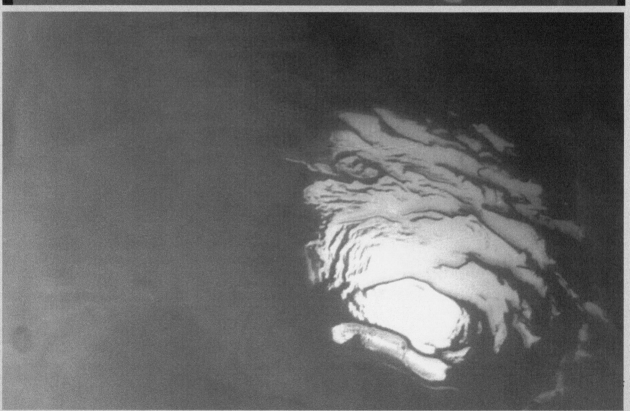

Decades of freely-available spacecraft imagery, combined with phenomenal advancements in computing technology have empowered citizen scientists and artists to revisit historic space missions. IAAA artist Don Davis explained some of the process: *"Although rudimentary digital image processing was used to bring out detail in low contrast images, especially from Mars, the end products of the missions were made available as photographic prints to space centres, publishers and interested individuals. Almost everyone would see such images in the form of reproductions in magazines or on television.*

Once space missions began creating online archives of their images and other data, opportunities arose for people to work with these types of images themselves. Don Davis continued: *"Until the mid-1990's the computer power needed to process images was a high end custom built capability. Home computers with software like Photoshop by then were surpassing the abilities of the systems which originally prepared many of the released space images. By 2010 a community of digital space image processors had emerged, each with their own emphasis and methods or style. Almost all of them have a working knowledge of the subject matter, and are working to bring out some aspect of the subject they find intriguing. As missions proceeded to place archives on line in near real time as well as in later refined products, many opportunities have arisen to work with grayscale and color photography from across the Solar System.*

Don's career as a space artist influenced his work in processing space images. He worked to process images into what they would look like to human visitors: *"This requires an effort to 'see past' the limitations of the cameras in relation to what the human eye would likely see. Dynamic range of the camera, or the ability to detect a wide brightness range at once, is a usual thing to factor in. The choice of colors to use is often an interesting effort depending on how well the photography performs at that task. Red, green and blue filtered exposures taken in rapid succession are excellent material to work with, as this method provides high color fidelity. When there are objects of known colors in the scene such as a color chart that helps a lot. Once the 'absolute' color, as seen in clear noon time daylight conditions on Earth, is reasonably achieved the 'ambient lighting' of the scene, as in the case of Mars and Venus, can be factored in to adjust the portrayal to local conditions."*

Spacecraft cameras often had limitations, where technology or budget made capturing a true colour image impossible. Don would then have to substitute missing colours, or force the scene somehow to look more natural. Don explained the limitations in the process: *"Sometimes such a portrayal is not an attractive option, as in a global view of Venus, which is a nearly featureless ball visually. Subtle aspects of original data can of course be manipulated to show things of interest beyond trying to imitate what an astronaut would obtain with a good digital camera or what one would see.*

What about images that were only ever returned in black and white? As it turns out, Don was still able to process these images into colour, though with a bit more work: *"With knowledge of what one is 'aiming for', and the willingness to perform at times 'invasive surgery' with an original photo a result can be achieved that would compare well with what a color camera would have obtained. In the case of the Spirit rover's last landscape a detailed color panorama was gathered from many color filtered pictures. A striking black and white panorama of wider angle NAVCAM pictures near Sunset provided what to me was the most dramatic lighting of any Martian scene ever obtained, and I used the color scene made under noon time conditions to color this carefully retouched black and white panorama, making sure every rock in the scene had its distinctive color."*

Such processing of spacecraft photos by dedicated enthusiasts has generally different methods used by each. With such individuals going through vast archives of image data and creating new products, fresh looks at our heritage of exploration are continually unearthed. This multiplicity of sources of space images in addition to the news releases of the government agencies encourages the circulation of inspiring visions.

(Facing page top) A raw image returned of the Martian south pole by Mariner 9. The image sufferes from dropouts, camera distortion and a dark point pattern (NASA). (Facing page bottom) Digital reprocessing and combining multiple monochrome images enables the original scene colours to be displayed (NASA).

ACKNOWLEDGEMENTS

This book would not have been possible without the gracious efforts of many individuals and organisations. I thank Alastair Reynolds for his forword. I thank Jon Ramer, Rachel Tillman, Cheryl Jacobs, Lauren Rikus, Meran Robinson, Trent and Julie McDougall for their help in editing this book. I would also like to thank David A. Hardy, Jon Ramer, William K. Hartmann, Don Davis, Mike Carroll, Dr Mark A. Garlick, Eileen McKeon Butt, Doug Forrest, Mark Pestana, Ray Cassell, Nicole Stott and Marilynn Flynn for kindly donating their artwork. I would also like to thank Anthony Wesley, Trent and Julie McDougall for their use of amateur astronomy images. I am grateful to Dan McGleese and Marc Rayman from NASA, who offered their time to this project. I am also grateful for Graziella Caprarelli and Mars Society Australia members Jon Clarke, the late David Willson and Siddarth Pandey for their assistance in this work. Ted Stryk provided valuable assistance in reprocessing rare Soviet-era Mars and Venus imagery.

I also gratefully acknowledge the efforts of NASA, JPL, the European Space Agency, JAXA, ISRO and the Chinese Academy of Sciences/ China National Space Application: The Science and Application Centre for Moon and Deep space Exploration for making spacecraft imagery available. Finally I would like to thank the astronomer, space pioneers and space artists who have gone before us and helped humanity imagine, and explore our neighbouring worlds. All artwork not credited in the caption is my own.

The following websites helped make this book possible (correct as of October 2018):

- jpl.nasa.gov (NASA space missions)

- esa.int (European Space Agency)

- global.jaxa.jp (Japan Aerospace Exploration Agency)

- marssociety.org (Mars Society)

- iaaa.org (International Association of Astronomical Artists)

- theartisticastronaut.com (Nicole Stott)

- astroart.org (David A. Hardy's site)

- markgarlick.com (Mark A. Garlick)

- black-cat-studios.com (Ron Miller)

- psi.edu/about/staff/hartmann/painting (Bill Hartmann)

- eileenmckeonbutt.com (Eileen Mckeon Butt)

- apollo-arts.com (Doug Forrest)

- pestanafineart.wordpress.com (Mark Pestana)

- latest.raycassel.com (Ray Cassel)

- tharsisworks.com (Marilynn Flynn)

- stock-space-images.com (Michael Carroll)

- donaldedavis.com (Don Davis)

- stevenhobbsphoto.com.au (Steven Hobbs)

FURTHER READING

Atkinson, N., Incredible Stories From Space, Page St Publishing, 2016.

Ball, R.S., Story of the Heavens, Cassell and Company, 1891.

Ball, R.S., Star-Land, , Cassell and Company, 1895.

Cattermole, P., Moore, P., Atlas of Venus, Cambridge University Press, 1997.

Chaiken, A., A Passion for Mars, Abrams, New York, 2008.

Chaiken, A., A Man on the Moon, Penguin Books, 1994.

Collins, M., Liftoff: The Story of America's Adventure in Space, Grove Press, 1988.

Cross, C.A., Moore, P., The Atlas of Mercury, Crown Publishers, 1977.

Fielder, G., Wilson, I., Volcanoes of the Earth, Moon and Mars, Elek Science, 1975.

Fimmel, R.O., Van Allen, J., Burgess, E., Pioneer: First to Jupiter, Saturn and Beyond, NASA SP: 446, 1980.

Fimmel, R.O., Colin, L., Burgess, E., Pioneer Venus, NASA SP: 461, 1983.

Greeley, R., Introduction to Planetary Geomorphology, Cambridge University Press, 2013.

Grindspoon, D.H., Venus Revealed, Helix Books, 1997.

Hanlon, M., The Worlds of Galileo, Constable Publishers, 2001.

Hanlon, M., The Real Mars. Constable and Robinson, 2004.

Hardy, D.A., Moore, P., The New Challenge of the Stars, Hutchinson Group, 1977.

Hardy, D.A., Visions of Space, Paper Tiger, 1989.

Hobbs, S.W., The Formation and Evolution of Gullies on Mars and Earth: A Complex Interplay Between Multiple Agent Processes, University of New South Wales, Canberra, 2014.

Huxley, J., Worlds Beyond Ours, Odham Books, 1968.

Jones, M., The New Moon Race, Rosenberg Publishing, 2009.

Kelly Beatty, J., Exploring the Solar System, National Geographic, 2001.

Ley, W., The Conquest of Space. Sidgwick and Jackson Ltd, 1951.

Lorenz, R., Mitton, J., Lifting Titan's Veil, Cambridge, 2002.

Lowell, P., Mars. Longmans, Green & Co., 1896.

Lowell, P., Mars and its Canals. Macmillan Co., 1906.

MacPherson, H., Modern Astronomy: Its Rise and Progress, Oxford University Press, 1928.

McNab, D., Younger, J., The Planets, BBC Worldwide, 1999.

Miller, R., Hartmann, W.K., Traveller's Guide to the Solar System, Workman Publishing Company, 1981.

Mitchell, O.M., Orbs of Heaven, George Routledge and Sons, 1850.

Moore, P., The Next Fifty Years in Space, Taplinger Publishing Company, 1976.

Morrison, D., Samz, J., Voyage to Jupiter, NASA SP: 446, 1980.

Mortillaro, N., Saturn: Exploring the Mystery of the Ringed Planet, Firefly Books, 2010.

National Commission on Space, Pioneering the Space Frontier, Bantam Books, 1986.

Olmstead, D., Mechanism of the Heavens, Thomas Nelson, 1850.

Preiss, B., The Planets, Bantam Books, 1985.

Proctor, R.A., The Poetry of Astronomy, Smith, Elder & Co., 1881.

Pyle, R., Destination Moon, Carlton Books, 2005.

Sagan, C., Mars: a New World to Explore, National Geographic, 1967.

Sartwell, F., Voyage to Venus: The Story of Mariner II, National Geographic, 1963.

Shelton, W.R., Man's Conquest of Space, National Geographic, 1974.

Slipher, E.C., New Light on the Changing Face of Mars, National Geographic, 1955.

Stern, S.A., Our Worlds, Cambridge University Press, 1999.

Stern, S.A., World Beyond, Cambridge University Press, 2002.

Strughold, H., The Green and Red Planet, Sidgwick and Jackson, 1954.

The Viking Lander Imager Team, The Martian Landscape, NASA SP: 425, 1978.

Vaucoulers, G., Physics of the Planet Mars, Faber and Faber Ltd, 1954.

Vita-Finzi, C., Fortes, D., Planetary Geology, Dunedin Academic Press, 2013.

Weaver, K.F., Voyage to the Planets, National Geographic, 1970.

9 780648 447603